Toward Integrated Water Resources Management in Armenia

DIRECTIONS IN DEVELOPMENT
Countries and Regions

Toward Integrated Water Resources Management in Armenia

Winston Yu, Rita E. Cestti, and Ju Young Lee

WORLD BANK GROUP

Contents

Boxes

Figures

Maps

Photos

Tables

Acknowledgments

This report was prepared by a team led by Winston Yu together with Rita Cestti, Ju Young Lee, Arusyak Alaverdyan, and Vahagn Tonoyan. Hrishi Prakash Patel provided support on the production of maps and geographic information systems. We are grateful to Dina Umali-Deininger, Jean-Michel Happi, and Henry Kerali for the valuable support, encouragement, and guidance to the team. Their recognition of the importance of this exercise helped to facilitate the team's engagement with its country counterparts. Thank you also to Satik Nairian and Nora Mirzoyan for the invaluable logistical assistance during our many visits to Armenia. Our often hectic meeting schedules went as smoothly as could have been asked for.

The authors benefited enormously from many discussions with colleagues within the World Bank, government of Armenia officials, and representatives from donor and nongovernmental organizations working in the Armenia water sector, including Zaruhi Tokhmakhyan, Arthur Kochnakyan, and Adriana Damianova (World Bank); Andranik Andreasyan, Garegin Bagramyan, Armen Harutyunyan, Karen Papoyan, Levon Vardanyan, Marzpet Kamalyan, Araksya Davoyan, Arthur Gevorgyan, Liana Margaryan, Vladimir Tadevosyan, Adibek Ghazaryan, Karen Grigoryan, and Tigran Ishkhanyan (government of Armenia); and Arevik Hovsepyan and Inessa Gabayan (other institutes working on the water sector). All of the close cooperation, generosity, and candid views we received have made this report possible and more meaningful.

Participants at a donor meeting on January 22, 2014 included Marina Vardanyan, Lusine Taslakyan, John Baker, Fukumori Daisuke, Ruzan Khojikyan, Armen Paghosyan, Bella Andriasyan, and Zara Chatinyan. This meeting offered a useful opportunity for all donors to share their past, present, and future work programs in the water sector. The team looks forward to engaging further with this group.

The authors also acknowledge constructive comments and suggestions from the following reviewers: Eileen Burke, William Rex, Nagaraja Rao Harshadeep, Ulrich Bartsch, Satoru Ueda, Marcus Wishart, Mark Lundell, and Aleksan Hovhannisyan. Their contributions have helped to enhance the quality of this report, refine and focus the key messages, and guide the team on the strategic dialogue forward on these important issues.

About the Authors

Winston Yu is a senior water resources specialist at the World Bank. He has extensive experience in technical and institutional problems in the water sector and has carried out a number of research and investment projects in developing countries. His special interests include river basin management tools, hydrologic modeling, flood forecasting and management, groundwater hydrogeology, international rivers, and climate change. Prior to joining the World Bank, Yu was a researcher at the Stockholm Environment Institute and served as a science and technology officer at the U.S. Department of State. He is also an adjunct professor at the School of Advanced International Studies at Johns Hopkins University, where he is associated with the Global Water Program. He received a PhD from Harvard University.

Rita E. Cestti is a senior water resources specialist at the World Bank. Her main areas of work include water resources management and development and climate change adaptation and safeguards. She has managed the identification, preparation, and supervision of a number of water-related, natural resources, environmental and disaster management projects in several countries, leading the preparation of economic sector works and implementation of technical assistance activities. She has also conducted extensive economic studies in the context of sector work and project analysis in water resources management and development. She holds a professional degree in civil engineering from the Pontificia Universidad Católica, Peru, and master's degrees in engineering administration and economics from The George Washington University, Washington, DC. In addition, she completed course work in economics at the PhD level, and is a registered professional civil engineer in Peru.

Ju Young Lee is a junior professional associate at the World Bank. Her World Bank experience includes investment projects in Armenia, Kosovo, the Kyrgyz Republic, Ukraine, and Uzbekistan, and research projects for the Aral Sea Basin (Central Asia) and the Drina River Basin (Bosnia and Herzegovina, Serbia and Montenegro). Prior to joining the Bank, she also conducted research into rainwater harvesting in Tanzania and volunteered in India for water supply and sanitation issues. Her interests include hydrologic modeling, groundwater hydro-geology, transboundary waters, integrated water resources management, and the water-energy-food nexus. She studied water resources engineering at both the graduate and undergraduate level and holds an MS degree from Stanford University and a BS from Columbia University.

.

Abbreviations, Currency, and Measurements

Abbreviations

ADB	Asian Development Bank
AFD	French Agency for Development
AMD	Armenian dram
ASHMS	Armenian State Hydrometeorological and Monitoring Service
BMO	basin management organization
EBRD	European Bank for Reconstruction and Development
EC	European Commission
EDB	Eurasian Development Bank
EU	European Union
FAO	Food and Agriculture Organization of the United Nations
GDP	gross domestic product
GEF	Global Environment Facility
GIZ	German Agency for International Cooperation
IFC	International Finance Corporation
IWRM	integrated water resources management
JICA	Japan International Cooperation Agency
KF	Kuwait Fund for Arab Economic Development
KfW	KfW Development Bank
MCC	Millennium Challenge Corporation
OECD	Organisation for Economic Co-operation and Development
OSCE	Organization for Security and Co-operation in Europe
RBMP	river basin management plan
SEI	State Environmental Inspectorate
Sida	Swedish International Development Cooperation Authority
SWCIS	State Water Cadastre Information System
UNDP	United Nations Development Programme
UNECE	United Nations Economic Commission for Europe
USAID	United States Agency for International Development
WRMA	Water Resources Management Agency

Currency Equivalents

(Exchange rate effective as of August 13, 2014)
US$1.00 = 415 AMD
AMD 1.00 = US$0.0024

Weights and Measures

Metric system

Overview

Integrated Water Resources Management Diagnostic

The proper management of water resources plays a key role in the socioeconomic development of Armenia. On average, Armenia has sufficient water resources. Taking into account all available water resources in the country, Armenia has sufficient resources to supply approximately 3,100 cubic meters per capita per year—well above the typically cited Falkenmark water stress indicator of 1,700 cubic meters per capita per year. These water resources are not evenly divided in space and time with significant seasonal and annual variability in river runoff. In order to address temporal variations in river runoff, the country has built 87 dams with a total capacity of 1.4 billion cubic meters. Most of these dams are single purpose, mainly for irrigation. Armenia also has considerable groundwater resources, which play an important role in the overall water balance. About 96 percent of the water used for drinking purposes and about 40 percent of water abstracted in the country comes from groundwater. Irrigation remains the largest consumptive user (figure O.1).

Agriculture in Armenia is heavily dependent on irrigation. More than 80 percent of the gross crop output is produced on irrigated lands. Returns are higher on irrigated lands. Water user associations play an important role in agricultural water management. Currently, there are 42 water user associations responsible for about 195,000 hectares (out of a total of 208,000 hectares of irrigable lands in Armenia). Since water user associations became operational, water supply has improved, the collection of water fees has increased, and there is an increasing conversion from low-value crops (e.g., wheat) to higher value crops (e.g., fruits and vegetables) (table O.1). However, water user associations are not yet financially sustainable and continue to depend on State subsidies. Finally, agricultural water management is still subject to various inefficiencies. This includes the widespread use of high-lift pump irrigation systems built during Soviet times but are now uneconomical due to high energy costs.

Domestic water consumption, which used to be the second-largest water user after irrigation, sharply decreased in the 1990s (figure O.2). This dramatic drop is attributed to the introduction of water metering and a volumetric billing system.

Figure O.1 Water Consumption by Sector, 1995–2012

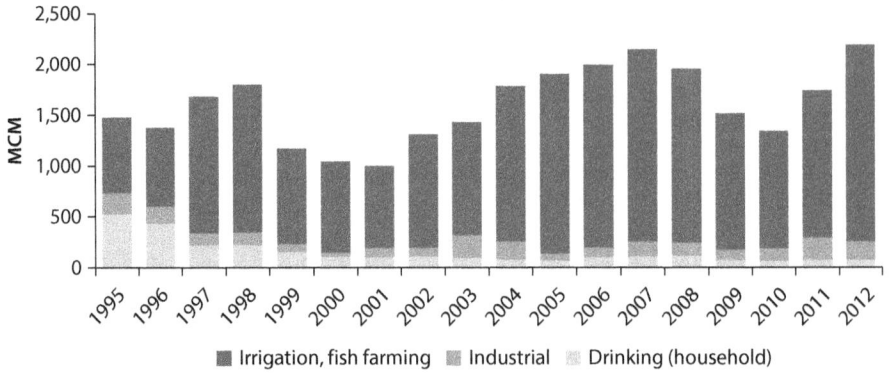

Source: National Statistical Service of Armenia.
Note: MCM = million cubic meters.

Table O.1 Improvements after the Operationalization of Water User Associations, 2004–2013

	2004	2005	2006	2007	2008	2009	2010	2011	2012	2013
Irrigated area (ha)	113,366	125,648	123,298	125,632	128,860	128,076	129,194	129,406	130,180	130,524
Collection (billion AMD)	2.51	2.89	2.95	3.10	3.44	3.22	3.56	3.77	4.03	4.44
Collection rate (%)	56	66	69	73	68	87	82	83	78	86
High-value crops (%)	65	71	74	78	79	79	80	84	87	88

Source: Project implementation unit data.
Note: AMD = Armenian drams. A billion is 1,000 million.

Figure O.2 Water Consumption for Domestic, 1995–2012

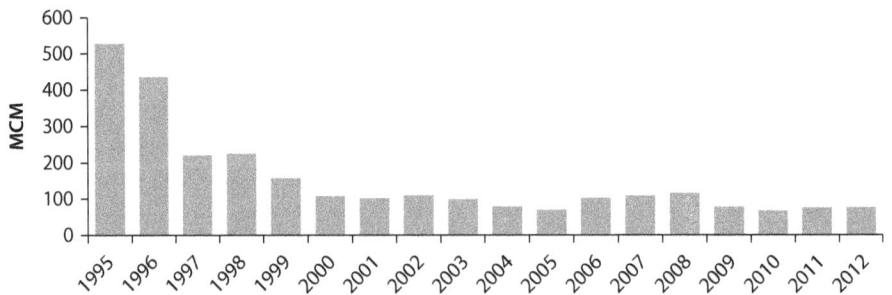

Source: National Statistical Service of Armenia.
Note: MCM = million cubic meters.

Over the past decade, water supply in Armenia has greatly improved with the increased use of public-private partnerships. This has shown success, particularly with improving water supply duration, water meter installment, and collection efficiency. Compliance with water quality requirements has also improved and energy consumption has, in most cases, been reduced. Although the collection rate is high, the tariff is still currently too low to provide sufficient funding to cover even routine operation and maintenance and investment costs. Moreover, while water supply has greatly progressed, sanitation has fallen behind. Wastewater collection and treatment systems are not sufficiently provided and operational.

Lake Sevan has environmental, economic, and social significance and is an important multipurpose water reservoir for irrigation, hydropower, and recreational uses. The level of Lake Sevan fell dramatically due to excessive use during the period from 1930 to the 1980s, resulting in serious environmental and ecological problems, including deterioration of water quality, destruction of natural habitats, and loss of biodiversity. Starting in the 1980s, programs to stabilize and raise the lake level were initiated. This includes the construction of the Arpa-Sevan and Vorotan-Arpa tunnels, transferring up to 250 and 165 million cubic meters (MCM), respectively, and outflow limits up to 170 MCM per year. As a result, the level of Lake Sevan has been steadily rising since 2001 (figure O.3). Overfishing continues to be a major problem in the lake.

Water resources also play a critical role in the energy sector. Armenia has great potential for hydropower from its mountains and fast-flowing rivers. There are two large cascades and a number of small hydropower plants. The total installed capacity from hydropower is 1,032 megawatts (figure O.4). The last decade has witnessed a major growth in the numbers of private small hydropower plants, spread throughout the country. As of 2012, there are 129 existing small hydropower plants with a capacity of 210 megawatts, and 75 more under construction with a capacity of 156 megawatts. Recent analysis finds that an additional 250–300 megawatts of generation is possible from small hydropower plants. Some have raised concerns regarding the impact of existing and future small hydropower plants on water resources and environmental sustainability.

Compared with other countries in the region, Armenia is highly vulnerable to climate change. Armenia shows high exposure, high sensitivity, and limited

Figure O.3   Water Releases from Lake Sevan and Lake Level, 1927–2011

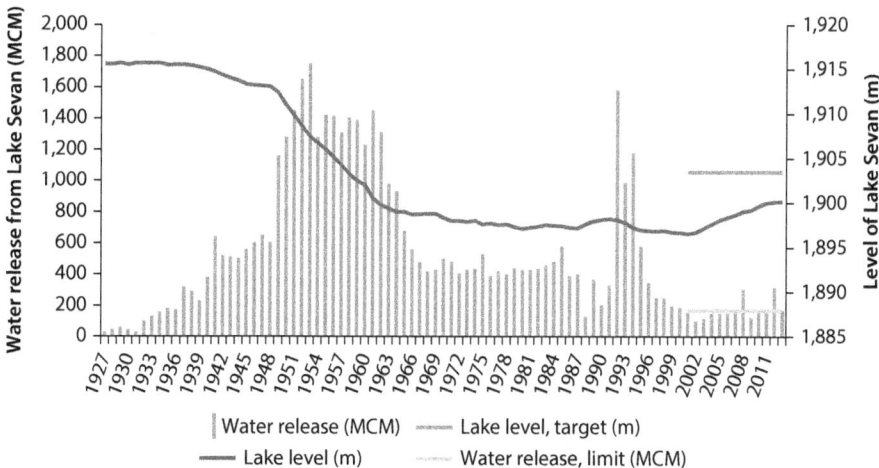

Source: State Committee on Water Economy.
Note: MCM = million cubic meters.

Figure O.4 Net Installed Capacity, 2005–10

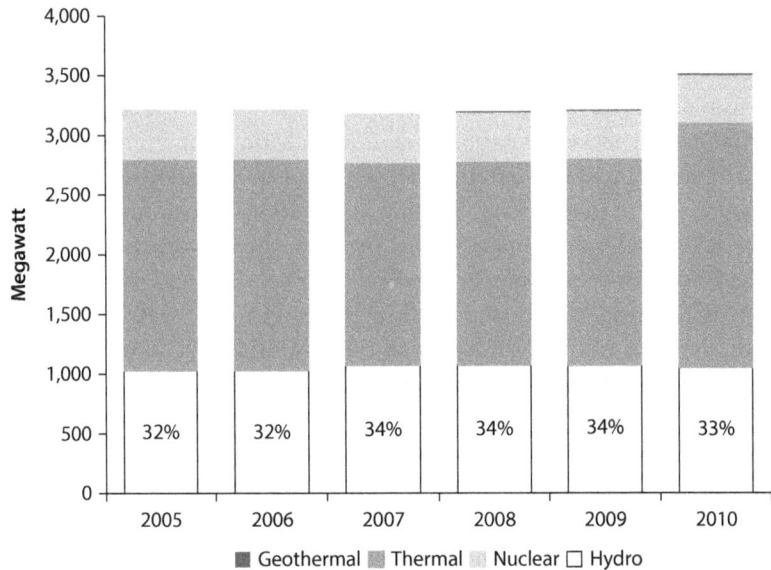

Source: United Nations 2010.

adaptive capacity to climate change. Future climate projections indicate continued increases in temperature and decreases in precipitation. The impacts of climate change will be particularly severe for Lake Sevan. In the agriculture sector, the most climate-sensitive sector, crop yields are predicted to decline and irrigation demands to increase with climate change. It is estimated that by 2030, yields of the main agricultural crops will decrease by 8–14 percent without adaptation. In order to maintain crop yields, substantially more irrigation will be needed. However, with overall water resources availability expected to decline, these demands may be difficult to fully meet in the future. A 25 percent reduction in river flow is projected to result in a 15–34 percent reduction in the productivity of irrigated cropland (average 24 percent). The energy sector will also be affected, as Armenia uses its rivers for hydropower generation and cooling water for nuclear and thermal power plants. Finally, climate change is likely to decrease water supply in transboundary basins.

A Decade of IWRM Reform

Over the last 10 years, Armenia has achieved significant legislative and institutional reforms in terms of water resources management and protection. Notable among these are the adoption of the updated Water Code in 2002, the Law on Water User Associations and Federations of Water User Associations in 2002, the Law on the Fundamental Provisions of the National Water Policy in 2005, and the Law on the National Water Program in 2006. These measures establish the principles and mechanisms needed to implement integrated water resources management (IWRM) in the country. In general, these laws are quite extensive and comprehensive in scope and serve as a strong foundation for planning and management in the water sector. The main agencies responsible for implementing IWRM are given in table O.2.

A system of tariffs and fees are used to regulate water uses (both consumptive and nonconsumptive) (table O.3). Current expenditures exceed what is collected through these various fees.

Further institutional strengthening is needed to fulfill the vision of this legislative framework. To date, many of the National Water Program measures have not been implemented or have been largely supported by international donors and through bilateral assistance (table O.4). Thus, technical capacity and the necessary internal budgets to implement these measures have been insufficient. The water resources management, monitoring, and compliance assurance organizations (Water Resources Management Agency [WRMA], Basin Management Organizations [BMOs], Armenia State Hydrometeorological and Monitoring Service, Environmental Impact Monitoring Center, Hydrogeological Monitoring Center, State Environmental Inspectorate) together receive annually around 500 million Armenian drams (US$1.2 million) for their water-related activities. It is estimated that about 1.7 billion drams (US$4.1 million) is required to fully and properly implement the tasks and responsibilities assigned to these agencies.

Moreover, second-generation reforms are needed, which include support to the decentralization process, strengthening the water permit system, strengthening the monitoring system, and broad-based capacity building on IWRM, particularly with respect to river basin planning. Building the capacity of the WRMA and BMOs will be critical in these regards.

This is now even more important in the context of emerging challenges in the water sector. These challenges includes continued deterioration of the country's monitoring network (both quantity and quality, for both groundwater and surface water), poor water resources planning (from the river basin perspective), continued weak enforcement under the water permit system (the main regulatory function), concerns over the multitude of water issues in the Ararat valley, increased concerns over transboundary issues, and increased needs for strategic development and management of surface water storage.

Emerging Challenges to IWRM

Strengthening Monitoring of Water Quantity and Quality

Obtaining reliable, timely, good-quality, and publicly available data on water quantity and quality are precursors to a properly functioning water management and planning system. Future investments cannot be fully prepared without a sufficient knowledge base on water resources in place. Moreover, day-to-day operations of the various water systems both for productive purposes (for example, irrigation, urban supply, environmental flows) and risk mitigation purposes (for example, flood warning) cannot be optimized without a robust near real-time

Table O.2 Main Institutions for IWRM in Armenia

	Management and protection of water resources	Regulation of tariffs	Management of water systems
Authorized agency	Water Resources Management Agency (WRMA)	Public Services Regulatory Commission (PSRC)	State Committee on Water Systems (SCWS)
Main functions	Monitoring and allocation of water resources, strategic management and protection of water resources	Regulation of tariffs for noncompetitive water supply and discharge services in drinking, household, and irrigation water sectors; protection of consumers' rights	Management of water systems under State ownership, support to establishment of water user associations and unions of water users, arrangement of tenders on management of water systems
Enforcement tools	Water use permits	Water system use permits	Management contract

Table O.3 Water Tariffs and Fees by Sector

Economic instrument	Beneficiary	Management objective	Sector
Abstraction fee	Water Resources Management Agency (WRMA)	Rational use and efficient allocation of water resources, ensuring minimum environmental flow	Drinking water (household), industrial, irrigation, fisheries sectors
Pollution fee	WRMA	Pollution reduction	Industry, urban wastewater supply, irrigation, fisheries sectors
Tariff	Private companies, local administrations, supply, and discharge companies	Sustainable water supply to population	Irrigation water supply, drinking water supply, and discharge
Fines and penalties	State Environmental Inspectorate (SEI)	Compliance with water use permit conditions, pollution reduction, ensuring minimum environmental flow	All entities holding water use permits

Source: Ministry of Nature Protection 2013.

Table O.4 Implementation Status of Short-Term Measures of the National Water Program

Issues and short-term measures	Implementation status[a]		
	1	2	3
Legal requirements			
1. Intersectoral harmonization and improvement of the existing legislation			
2. Establishment of an interagency standing commission within the National Water Council (NWC) to discuss amendments to be made to the legal acts			
Institutional development			
3. Review and implementation of developed recommendations related to overlaps and gaps in the roles and responsibilities			
4. Adjustment and improvement of the mechanisms for interagency cooperation and coordination by the NWC			
5. Development of a program for institutional development of the bain management organizations (BMOs)			
Water resources management needs			
6. Development and testing of a pilot monitoring system in one basin management area			
7. Development of a monitoring strategy and a national program			
8. Reestablishment of the groundwater resources monitoring system in Armenia			
9. Improvement of the existing water use permit regulations, and establishment of criteria for priority of water use application			
10. Development of criteria and guidelines for environmental impact assessment as part of the water use permit application process			
11. Development and implementation of a short-term program for the State Water Cadastre			
12. Ensuring public awareness and participation in the planning and management of water resources at the national and basin levels			
13. Development and implementation of strategies for establishment of basin public councils, and technical capacity building			
14. Implementation and continuous monitoring and assessment of the National Water Program			
15. Establishment of a monitoring system for the program implementation			
16. Capacity building in the Water Resources Management Agency (WRMA) and basin management organizations for integrated water resources management (IWRM)			
17. Development of a pilot river basin management plan (RBMP) and identification of information needs for one basin management area			
18. Review and improvement of the programs of measures for restoration, protection, reproduction, and use of the Lake Sevan ecosystem			
19. Clarification of up-to-date characteristics of water resources and water reserve components			
20. Adjustment and introduction of an international methodology for determination of norms for the limitation of impacts on water resources			
21. Development of a methodology for determination of aquatic ecosystem protection zones			
22. Development and implementation of programs for use of previously drained agricultural lands in the Ararat valley			
23. Implementation of works provided for under the program for reservoir construction			
24. Development of a strategy for water quality management			
25. Review and improvement of the existing approaches to spatial planning			
26. Development of a program for management of transboundary water resources			
Water systems management needs			
27. Study of water supply and wastewater collection services and implementation of programs to improve the provided services			

table continues next page

Table O.4 Implementation Status of Short-Term Measures of the National Water Program *(continued)*

	Implementation status[a]		
Issues and short-term measures	*1*	*2*	*3*
28. Development of programs aimed at enhancing the measures for the safety of hydrotechnical structures and reliability of operations			
29. Clarification of responsibilities for operation and protection of hydrotechnical structures of State significance			

Note: NWC = National Water Council.
a. Implementation status: 1 = Started; 2 = Progress; 3 = Completed.

monitoring network. Finally, management of the overall resource sustainability (for example, through permitting) and various competing pressures is only possible when data are being monitored over time and resource assessments updated regularly. In Armenia, there are several different agencies with responsibility for water monitoring (both quantity and quality, both surface water and groundwater), as shown in table O.5.

The current monitoring system (table O.6) is quite weak and needs substantial investment (both in terms of hardware and human capital). Since Soviet days, very little investment has been devoted to strengthening the monitoring infrastructure. To enhance the current monitoring system, a comprehensive view must be taken. Over the last decade, investments in monitoring have been done in a piecemeal manner (a piece of equipment here, a piece of equipment there) with financing from outside donors. In most cases, the numbers of monitoring points could be expanded, the technologies used modernized (for example, through greater use of automated readers or real-time telemetry), and new approaches to data collection, verification, and management applied. Sharing of data among different agencies and access to data by the public (through department websites) also remains very limited. Some degree of harmonization across the various departments responsible for monitoring is needed.

Weakness in River Basin Management Planning

River basin management planning needs to be improved, and a strategic vision is required for IWRM in each basin in the country. Despite the various initiatives and multiyear efforts supported by the donor community, the water sector in Armenia still faces many challenges with respect to river basin management planning due to weak capacity and inadequate information and analytical tools. The skills and data needed to carry out modeling and planning work are not yet available within the BMOs. The current river basin planning model relies heavily on the European Union Water Framework Directive and focuses primarily on achieving good ecological status of water bodies. Broad intersectoral planning that takes into account water, agriculture, energy, and environment linkages is not sufficiently developed. Several draft river basin management plans (RBMPs) have been already developed or are in the process of development (Debed, Aghstev, Marmarik, Vorotan, Meghriget, Arpa, Akhuryan, Metsamor river basins

Table O.5 Water Monitoring Institutions in Armenia

Monitoring function	Responsible agency	Ministry
Surface water quantity	Armenia State Hydrometeorological and Monitoring Service	Emergency Situations
Surface water quality	Environmental Impact Monitoring Center	Nature Protection
Groundwater quantity and quality	Hydrogeological Monitoring Center	Nature Protection
Drinking water sources and quality	State Health Inspectorate	Health Care
Water use and pollution discharge	State Environmental Inspectorate	Nature Protection

Table O.6 Summary Information on Surface Water and Groundwater Monitoring Points

Basin	Area (km²)	Surface water quantity gauging stations No.	Surface water quantity gauging stations Km² for 1 station	Surface water quality sampling points No.	Surface water quality sampling points Km² for 1 station	Groundwater springs and wells
Akhuryan	5,044	17	297	14	360	14
Ararat	4,460	13	319	16	279	8
Northern	7,068	23	307	25	283	39
Sevan	4,806	14	339	22	216	3
Hrazdan	3,881	16	243	33	118	1
Southern	4,484	9	498	21	213	8
Total	**29,743**	**92**	**334**	**131**	**245**	**73**

Source: European Union 2011.

(map O.1), but the government has yet to officially adopt, fund, or implement any of these plans. Government endorsement of such plans is needed to ensure that all levels of government have a consistent approach to water management and clear prioritization of future investments. Analysis and knowledge of what would be the best allocation (both in terms of economics and efficiency) for the different water users in each basin is needed. This is despite the fact that water permit and allocation decisions are routinely being made. Currently, the planning of irrigation, water supply, and hydropower investment programs, which are managed at the central level, has limited relationship with the RBMPs. Thus, a clear disconnect exists between the basin plans and sector programs and budgets. Moving forward, the government will need to invest budgetary resources in these multidepartmental basin planning efforts.

Strengthening the Water Permit System

Water use permits are one of the key tools for management and allocation of water resources in the country. Improved implementation of the water use permit system is still constrained by deficiencies in permitting regulations, insufficient cooperation among agencies in the processes of issuance of permits and the assurance of compliance with permit conditions, and capabilities and resources of agencies and their staff. The WRMA is the principal agency responsible for

Map O.1 Coverage of RBMPs in Armenia

Rivers

Lakes and reservoirs

BMOs

Sub-basins

RBMP Status

No plan

Ongoing (EU EPIRB)

Draft Plan (EU/UNDP/GEF)

Marmarik Basin Financial and Economic Studies (UNECE)

Environmental Action Plan (WB/UNDP)

Draft Plan (UNDP/GEF)

Ongoing (USAID)

Note: A full-color version of this map may be viewed at http://www.issuu.com/world.bank.publications/docs/9781464803352. EU = European Union, EPIRB = Environmental protection of international river basins, UNDP = United Nations Development Programme, GEF = Global Environment Facility, UNECE = United Nations Economic Commission for Europe, WB = World Bank, USAID = U.S. Agency for International Development.

issuing water permits. Through decentralization, this function (in the long term) is expected to be devolved to the BMOs.

Ensuring compliance with water permits is currently hampered by insufficient resources and weak agency capacity. Currently, compliance involves a monitoring function (WRMA) and an enforcement action function (SEI). These roles and responsibilities have been separated. Though this separation is advantageous, greater cooperation and coordination (perhaps legislated) on inspection and enforcement is needed between the WRMA and the SEI. In the future, compliance history could be made a more explicit part of the permitting process and greater compliance promotion (and more reliance on self-monitoring) undertaken by the government. Refining the permitting procedures for small, medium, and large water uses and pollution discharges may enhance the permit process, including establishment of a limit of withdrawal and pollution discharge below which a water use permit is not required. Finally, greater public participation in the permitting process may be envisioned to provide greater transparency.

The Future of Ararat Valley

The Ararat valley is the largest agriculture and fish farming zone and has strategic importance to the Armenian economy. The Ararat valley is rich with high-quality artesian groundwater, which is suitable for drinking purposes without additional treatment and comprises a strategic reserve of drinking water for the country. This resource has historically been used for drinking and irrigation purposes. Since 2006, a large number of fish farms have been established in the Ararat valley due to the rich supply of artesian groundwater of high quality and low cost, and have become one of the major water users. Fish production was included in the list of priority development programs in 2008, and thus more water use permits were issued for fish farms, exceeding the renewable level of groundwater resources. In 2013, groundwater use by fish farms alone exceeded the sustainable level, and the total groundwater use by all sectors in Ararat Valley was 1.6 times the level (figure O.5).

As a result of overissuance of water user permits and overabstraction of groundwater resources, artesian groundwater resources have sharply declined and the artesian groundwater zone has decreased (map O.2). Between 1983 and 2013, piezometric levels decreased on average by 6–9 meters, sometimes by as much as 15 meters. Well discharges have reduced by 6–200 liters per second. The artesian zone in the valley has also significantly reduced. The area with positive pressure decreased from 32,760 hectares in 1983 to 10,706 hectares in 2013.

This is causing conflicts with other artesian groundwater users—irrigation, domestic, industrial, and cooling waters. As the artesian area has reduced in the Ararat valley, the number of communities using artesian wells for irrigation and domestic water supplies has decreased from 44 in 1983 to 13 in 2013. Due to the reduced discharges of the Sevjur-Aknalich springs, the Armenian (Metsamor) nuclear power plant can take only half of its water requirement. Moreover, the fish farms' excessive discharge into the agriculture drainage system is problematic.

Figure O.5 Discharge of Operating Wells in Ararat Valley, 2007 and 2013

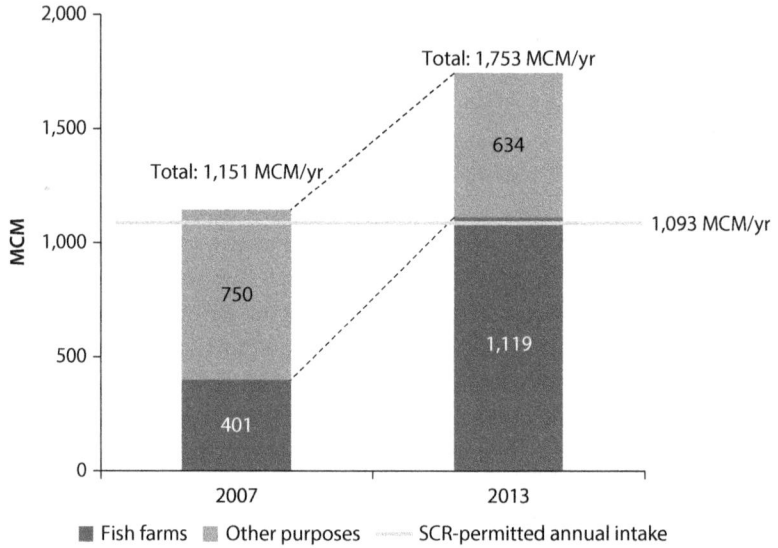

Source: USAID 2014.
Note: "Other purposes" include irrigation, drinking, and industrial water uses; MCM = million cubic meters.

Map O.2 Observations on Changes of Groundwater Levels and Pressure Zones in Ararat Valley

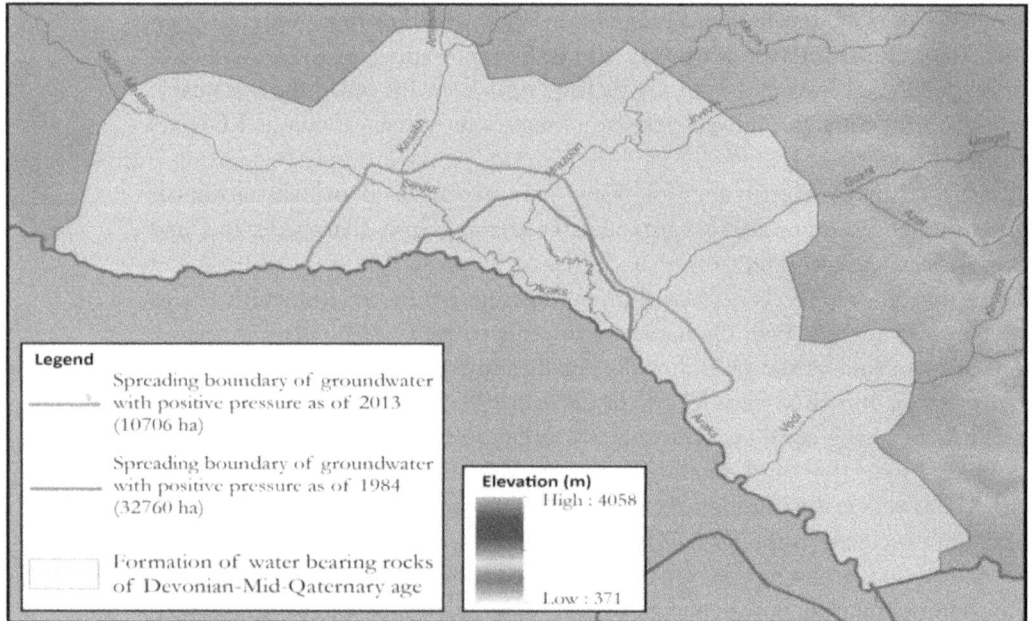

Source: USAID 2014.
Note: A full-color version of this map may be viewed at http://www.issuu.com/world.bank.publications/docs/9781464803352.

Toward Integrated Water Resources Management in Armenia • http://dx.doi.org/10.1596/978-1-4648-0335-2

Recognizing the growing concerns about water resources in the Ararat valley, several measures were adopted by the government, including stricter regulation over water use permitting and permit-enforcing processes and adjustments to the abstraction fees. The government promotes semiclosed water recycling system to fish farms, but it is not being widely adopted by fish farms for technical and financial issues. While short-term measures to restore and conserve artesian groundwater are being taken, coordinated action across a variety of departments responsible is urgently needed.

Transboundary Water Resources Issues

For Armenia, the transboundary nature of many of the rivers in the country creates a level of water insecurity. Important transboundary rivers include the Kura and Araks.[1] The Kura basin is shared with Azerbaijan, Georgia, and Turkey, and the Araks basin is shared with Azerbaijan, the Islamic Republic of Iran, and Turkey. Major proposed water infrastructure by Turkey (for irrigation, water supply, and hydropower purposes) is a major concern for the government of Armenia because of the expected flow impacts. The government has expressed willingness to collaborate with Turkey on the construction of a joint multipurpose dam on the Araks River along the Armenia-Turkey border (Surmalu dam), for which a joint technical concept has been prepared.

Deterioration of water quality in transboundary rivers is also a concern, for example, due to nonpoint source pollution from agriculture and livestock activities in the Araks and Akhuryan Rivers. Mining is also problematic as it relates to shared aquifers, such as the Aghstev-Tavush and Pambak-Debed aquifers. In addition to transboundary rivers and groundwater, there are important transboundary ecosystems shared by Armenia and Turkey in the Araks/Aras River valley. The Araks/Aras valley harbors several natural and artificial wetlands that provide important nesting areas for water birds.

Lack of formal cooperation between all the riparian countries and lack of a legal framework for transboundary cooperation are major limitations to making progress on this front. Most of the existing bilateral agreements between Armenia and its riparian countries, particularly those concluded with the Islamic Republic of Iran and Turkey, relate to water allocation. They may need to be revised to take into account water protection considerations. Existing agreements are silent with regard to transboundary groundwater issues. Implementation of bilateral agreements between Armenia and Turkey remains deficient. Though there are existing arrangements for the management of transboundary waters, the formal role of the WRMA in this regard is not properly addressed in the current legal framework.

Building Water Storage Capacity

Storage plays an important strategic role in the regulation of variable surface runoff in the country. This is critical for the irrigation, water supply, and energy subsectors, particularly in the semiarid regions where rapidly growing populations are facing depletion of groundwater resources. The country has built

87 dams, with a total capacity of 1.4 billion cubic meters. On average, the per capita storage capacity of Armenia is about 450 cubic meters, which is considered low for a semiarid country. In comparison to its neighboring countries, Armenian per capita storage is similar to that of the Islamic Republic of Iran, and represents less than 20 percent of the storage capacity of Azerbaijan and Turkey and less than 60 percent of the storage capacity in Georgia (figure O.6).

According to the Armenian Water Design Institute, there are 157 potential reservoirs at various stages of construction, design, or planning (table O.7). Most of the designs were completed during Soviet times. The overall storage capacity of these reservoirs is 1.72 billion cubic meters.

To move forward, a strategic plan for the development of priority reservoirs in Armenia is needed that addresses economic, financial, environmental, and social dimensions. Many of the earlier master plans were developed during the Soviet era and require updating and revisiting, especially with respect to their current technical and economic viability. Three key issues that will also need to be considered during the feasibility studies of the priority dams are climate change and transboundary impacts. First, with regard to climate

Figure O.6 Per Capita Storage Capacity in Armenia Compared to Its Neighbors and Other Countries

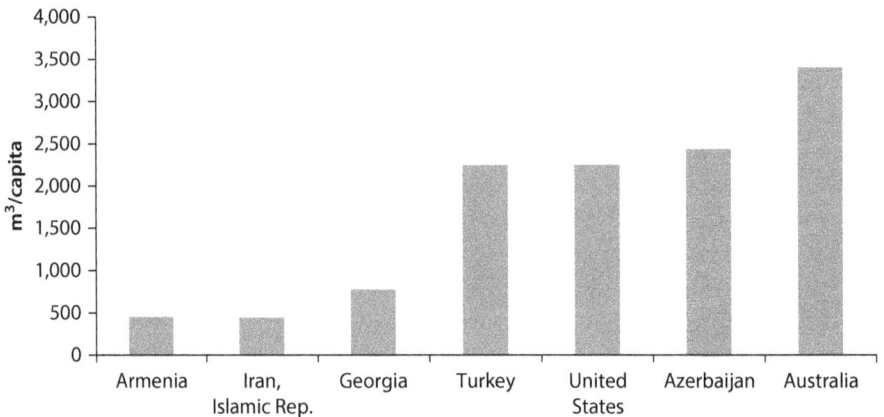

Source: FAO Aquastat database. Based on 2013 data.

Table O.7 Status of Reservoirs in Armenia

Status of reservoirs	Quantity	Storage volume (MCM)
Construction not completed	9	185.4
Designed (different stages of design)	23	733.2
Studied preliminary	67	452.8
Planned, but not studied	60	345.9
Total	157	1,717.3

Source: Water Design Institute of Armenia 2014.
Note: MCM = million cubic meters.

change, as the climate and hydrology have experienced changes since the investments were designed, it is important that the updated feasibility studies include these considerations. Second, with regard to transboundary impacts, as most of the rivers in Armenia are shared with neighboring countries down-stream, country impacts would need to be analyzed. Third, these large invest-ments should also be considered and analyzed within the context of overall river basin planning.

In addition, an overall financing strategy to support the proposed investments is needed. During the past few years, the government of Armenia has tried to mobilize external funding for completing the construction of unfinished reser-voirs and for updating the feasibility studies of those already designed reservoirs. Economic and financial costs and benefits need to be reassessed as well as the integrity of the existing works.

Recommendations

More actions and investment are clearly needed to fully realize the original vision as laid out in the Water Code and subsequent legislation. With the additional pressures and concerns described in the previous paragraphs, more effort is needed to ensure Armenia's future water security. Table O.8 synthesizes the recommendations made in this report and table O.9 gives some suggested areas where additional financing (and potential additional analytical support) would be required.

Table O.8 Synthesis of Report Recommendations

Issues	Recommendations
Financial sustainability for IWRM	• Some revision of existing tariff and fee structures may be required • Enhanced budgets to fulfill the mandates of the various institutions given in the existing legislative framework
Weak institutional (capacity) framework	• Continued skills and capacity development of water resource management institutions (particularly WRMA, BMOs, and water users associations) • Relative responsibilities of various actors need to shift toward greater focus on management
Need for second generation of reforms	• Completion of measures identified in the National Water Program (NWP) • Establishment of Secretariat under the National Water Council (NWC) to monitor and coordinate NWP recommendations and measures
Weak monitoring of water quantity and quality	• Investment in monitoring hardware (both quantity and quality) and staff skills development • Comprehensive review of overall monitoring network and future monitoring needs • Strengthening of public access to water-related data (i.e., revitalize the State Water Cadastre Information System) • Some harmonization across various departments and clarification of roles and responsibilities in monitoring
Weak river basin planning	• Development of skills and capacity (within WRMA and BMOs) for broad river basin planning (with focus on intersectoral concerns and investment planning) • BMOs need to take a more active role during the planning process. • Government allocation of budget resources to river basin planning efforts • Government endorsement of existing adequate river basin plans • Enhance economic considerations when preparing river basin management plans.

table continues next page

Table O.8 Synthesis of Report Recommendations *(continued)*

Issues	*Recommendations*
Weak implementation and administration of water permit system	• Governance and transparency issues need to be brought more forcefully. • Enhance cooperation among relevant agencies involved with issuance and compliance of permits • Development of skills for compliance assurance • Government allocation of budget resources to the permitting process • Inclusion of compliance history in permitting process • Greater promotion of self-monitoring • Refinement to permitting procedures for different water use levels • Enhance public participation in the permitting process
Growing water resource concerns in Ararat Valley	• Revisit the water permitting allocations in Ararat Valley • Some further revision of abstraction fees may be needed • Establishment of coordinating mechanism across several departments (e.g., SCWE, Ministry of Agriculture) to monitor status of Ararat Valley • Identification of affordable and economical technologies to reduce water use in fisheries
Growing transboundary water resource concerns	• The formal role for WRMA in transboundary management to be clarified • Revitalize the Armenian Commission on Transboundary Water Resources to more proactively engage in dialogue with its riparian neighbors
Insufficient water storage capacity	• Updating of storage master plans (in the context of river basin plans) to address economic, financial, environmental, and social dimensions. • Development of overall financing strategy for proposed storage investments
Weak donor coordination	• Mechanism needed to coordinate various donors on assistance in the water sector

Table O.9 Recommendations for Investment and Technical Assistance

Investment and technical assistance requirements	*Client*	*Comments*
Strengthening of overall water resources monitoring (including groundwater)	Ministry of Nature Protection	Given the current state of monitoring equipment in the field and the overlapping institutional responsibilities, harmonization and investment is needed. This would include investment in new technologies (both for quality and quantity) and capacity building of various agencies on quality assurance, quality control, data acquisition and storage, etc. Improved groundwater monitoring will be critical. This would support compliance with the European Union Water Framework Directive. A technical audit would be needed to assess the specific requirements, level of investment, and institutional strengthening needed.
Master planning of storage	Ministry of Territorial Administration	Technical assistance is needed to update feasibility studies for the individual reservoirs identified. A larger strategic evaluation and prioritization of all the numerous reservoir proposals is needed. This would look at the full range of economic, financial, environmental, and social issues and provide a framework for future analysis.
Comprehensive development program for Ararat valley	Ministry of Agriculture, Ministry of Nature Protection, Ministry of Territorial Administration	A comprehensive investment project is needed to address the many problems in the Ararat valley. A specific investment in this realm could provide an opportunity (and mechanism) for several ministries to work together. Investments specifically could be in groundwater, fish recycling technologies, drainage improvements, agriculture support, groundwater monitoring, etc.
Institutional strengthening of IWRM	Ministry of Nature Protection	Technical assistance is needed to help build the capacity of the primary IWRM agencies, particularly the WRMA and BMOs. The focus can be on strengthening existing river basin plans, strengthening the water permit process, twinning engagements with international partners on IWRM, etc.

Note

1. Alternative names for the rivers in this section include Kur, Kura (Georgia and Turkey), Mtkvari (Azerbaijan); Araks, Aras (the Islamic Republic of Iran and Turkey), Araz (Azerbaijan); Debed, Dobeda Chay (Georgia); Aghstev, Akstafe (Azerbaijan); Akhuryan, Arpaçay (Turkey); Vorotan, Bargyushad (Azerbaijan); Arpa, Arpa Chay (Azerbaijan).

References

European Union. 2011. *European Neighborhood Policy Instrument: Shared Environmental Information Systems (ENPI-SEIS), Armenia Country Report.*

FAO (Food and Agriculture Organization of the United Nations) Aquastat (database). Rome, Italy. http://www.fao.org/nr/water/aquastat/main/index.stm.

Ministry of Nature Protection. 2013. *Feasibility of the Master Plan for Integrated Water Resources Management in the Six Water Basin Management Areas of Armenia.* Report funded by SOFINEX and prepared by SHER Ingénieurs-Conseils.

United Nations. 2010. *Electricity Profiles (2005–10) for Armenia.* United Nations Statistics Division.

USAID. 2014. *Assessment Study of Groundwater Resources of the Ararat Valley.* Final report, prepared under USAID Clean Energy and Water Program.

CHAPTER 1

Introduction

Armenia is a small, landlocked country located in the southern Caucasus region. Its neighbors are Azerbaijan, Georgia, the Islamic Republic of Iran, and Turkey. It is a mountainous country with 75 percent of its land at higher than 1,500 meters above sea level. The average annual precipitation is 594 mm and the climate is considered semiarid and arid. The population is estimated at 3 million and has declined in recent years. Over the past two decades, the performance of the Armenian economy has ranged from a real gross domestic product (GDP) contraction of 42 percent following the collapse of the former Soviet Union to sustained annual growth rates over 10 percent between 2001 and 2008. In recent years, growth has been driven mainly by the mining sector and a strong recovery of agriculture (after a 16 percent drop in 2010, agriculture GDP grew at 14 and 9 percent in 2011 and 2012, respectively). In 2012, GDP was US$ 10 billion (4,000 billion Armenian drams) total and US$ 3,338 (1.3 million Armenian drams) per capita. Further macroeconomic details are given in appendix A.

The proper management of water resources plays a key role in the socioeconomic development of Armenia. About 80 percent of the country's crops are irrigated, with agriculture accounting for 15 percent of GDP. Net income per hectare, in general, is higher on irrigated lands. Hydropower accounts for 40 percent of total electricity production. Groundwater is the source of 96 percent of drinking water. Thus, it is no surprise that availability of water resources (figure 1.1) and its management are important determinants of the country's overall macroeconomic performance.

The government of Armenia recognizes the importance of integrated water resources management (IWRM), and toward this end has introduced over the last decade major institutional and policy reforms. Following the engagement of the World Bank in the early 2000s, the government initiated a targeted program of activities to strengthen management of the water sector and revised the legal and institutional framework. These were incorporated and adopted in the Water Code (2002) and subsequent National Water Policy (2005) and National Water Program (2006). These provide the legislative foundation and framework (and

Figure 1.1 Precipitation versus GDP

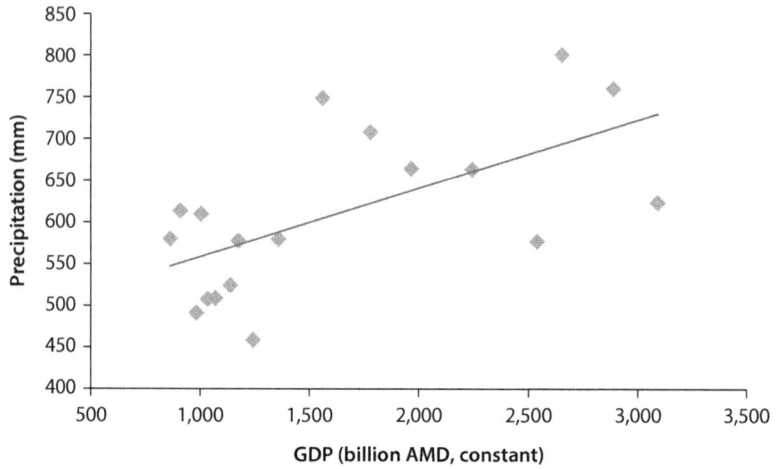

Source: World Bank Database and University of East Anglia's Climate Research Unit Database.
Note: Relationship is statistically significant ($t = 3.19$).

concomitant institutional bodies and processes) for ensuring the sustainable management and development of water resources in the country.

Despite this, further strengthening is needed to fulfill the vision of this legislative framework. In particular, many of the newly created institutions for IWRM—for example, the Water Resources Management Agency (WRMA) and the basin management organizations (BMOs)—require substantial technical and resource support. This is now even more important in the context of emerging challenges in the water sector. These include continued deterioration of the country's water monitoring network (both quantity and quality), increased concerns over transboundary issues, continued weak enforcement under the water permit system (the main regulatory function), unsustainable water usage in the important Ararat valley, and weak water resources planning (from the river basin perspective).

This report is organized in six chapters. Chapter 2 provides a diagnostic of water resources management in the country, focusing on critical water-using subsectors (for example, irrigation, domestic water, and environment) and future pressures. This chapter also includes a review of the impacts of climate change on the sector and the role of water in the energy sector. Chapter 3 provides a summary and review of the last decade of reform with IWRM and assesses the needs for continued strengthening of the policy framework to fully realize these achievements. The details of the various laws and institutions created are given in this chapter. Chapter 4 identifies some key emerging challenges for IWRM. These are areas that were identified after discussions with a wide range of stakeholders (including international donors and bilateral agencies). Chapter 5 presents the water engagement areas for the donor community (including the World Bank) over the last decade. Finally, chapter 6 concludes with some recommendations for further action and areas of potential investment and technical assistance support.

CHAPTER 2

Integrated Water Resources Management Diagnostic

Assessment of the Water Resources Baseline

On average, Armenia has sufficient water resources. Taking into account all available water resources in the country, Armenia has sufficient resources to supply approximately 3,100 cubic meters per capita per year,[1] well above the typically cited Falkenmark water stress indicator of 1,700 cubic meters per capita per year (Falkenmark 1989). All the rivers in Armenia are tributaries of the Araks and Kura Rivers. Most rivers are small, rapid, and fed by melting snow, springs, and groundwater. The overall river flow (originating within the country) has been estimated at 6.8 billion cubic meters (table 2.1) (USAID 2008b). This is in part driven by the estimated 16.7 billion cubic meters of precipitation, with less than 10.8 billion cubic meters lost by evaporation (USAID 2008b). An available 1.19 billion cubic meters originates from outside the country via the transboundary Araks and Akhuryan Rivers. Groundwater contributes an estimated 4 billion cubic meters. Note that there are discrepancies with regard to this baseline water balance (see appendix B) across various reported sources. Map 2.1 shows basin management organizations (BMOs) and river basins in Armenia.

These water resources are not evenly divided in space and time. Water resources are stressed, particularly in the densely populated Hrazdan River basin in the central part of the country (figure 2.1) (Ministry of Nature Protection 2010).

There is also significant seasonal and annual variability in river runoff, including frequent droughts and risk of flooding in the spring, when about 55 percent of total annual runoff occurs during the peak snow melting period (figure 2.2). The ratio of maximum to minimum flow can reach 10:1 (Ministry of Nature Protection 2010). For instance, the long-term (1953–2012) inflows into the Akhuryan reservoir are shown in figure 2.3. The coefficient of variation on the annual flows is 24 percent.

In order to address temporal variations in river runoff, the country has built 87 dams with a total capacity of 1.4 billion cubic meters. Most of these dams are

Table 2.1 Basin Management Organizations (BMOs) and River Basins in Armenia

BMO	River basin	Area (km²)	River flow (MCM/yr)
Northern BMO	Debed	3,895	1,203
	Aghstev	2,480	445
	Kura tributaries	810	199
Hrazdan BMO	Kasakh	1,480	329
	Hrazdan	2,565	733
Sevan BMO	Lake Sevan	4,750	265
Ararat BMO	Azat	952	232
	Vedi	998	110
	Arpa	2,301	764
Akhuryan BMO	Akhuryan	2,784	391
	Metsamor (Sevjur)	2,240	711
Southern BMO	Vorotan	2,476	725
	Voghji	1,341	502
	Meghriget	664	166
Total			**6,775**

Source: USAID 2008b.
Note: MCM = million cubic meters.

single purpose, mainly for irrigation. Thirty-five reservoirs have capacities greater than 1 million cubic meters (MCM), and three have capacities greater than 100 MCM.[2] There are 9 incomplete dams, 28 dams at the design stage, and a further 67 dams for which feasibility studies have been undertaken that were planned or prepared during the Soviet era (Ueda 2012). For the government of Armenia, the highest-priority dams for irrigation expansion and conversion from pump to gravity schemes are the Kaps, Vedi, Yeghvard, and Selav-Mastara. These are currently being financed (for prefeasibility studies and designs) or considered by several international donors. Lake Sevan, the largest freshwater body in Armenia, is another important multipurpose water reservoir for irrigation, hydropower, and recreational uses.

Armenia also has considerable groundwater resources, which play an important role in the overall water balance. About 96 percent of the water used for drinking purposes and about 40 percent of water abstracted in the country comes from groundwater (figure 2.4) (ADB 2011).

At present, the knowledge on availability and quality of groundwater resources in the country is limited due to the lack of monitoring. After the collapse of the Soviet Union, groundwater monitoring stopped for over 20 years and has only restarted in the last 4–5 years. In the last nationwide assessment of groundwater resources in the 1980s, total groundwater resources were estimated to be 4.0 billion cubic meters per year, which included 1.6 billion cubic meters of spring flow, 1.4 billion cubic meters of drainage flow, and 1.0 billion cubic meters of deep flow (table 2.2) (USAID 2008b). In the critical Ararat valley, deep groundwater resources are estimated to be about 1.8 billion cubic meters

Map 2.1 Basin Management Organizations and River Basins in Armenia

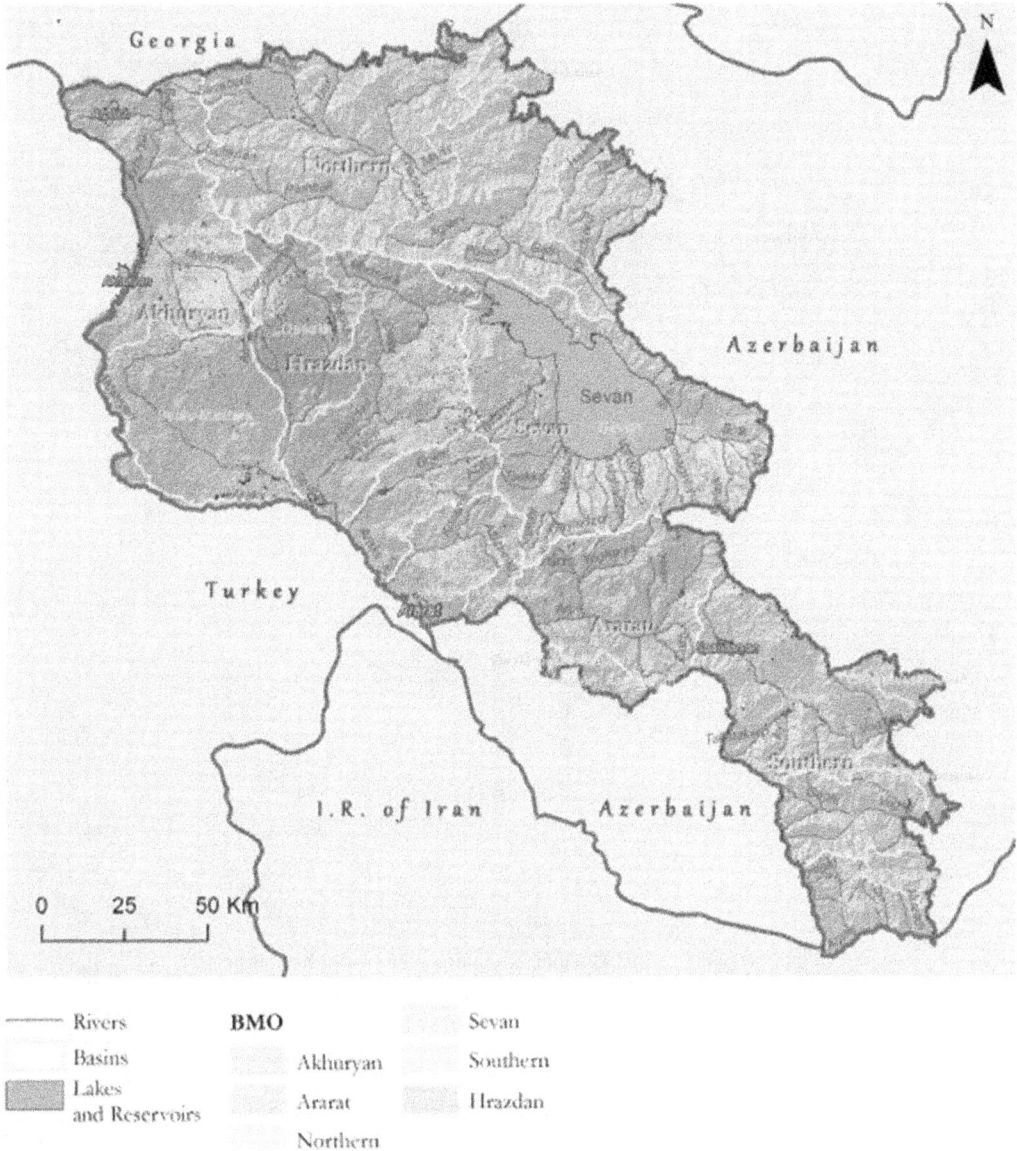

Rivers

Basins

Lakes
and Reservoirs

BMO

Akhuryan

Ararat

Northern

Sevan

Southern

Hrazdan

Source: USAID 2008b.
Note: A full-color version of this map may be viewed at http://www.issuu.com/world.bank.publications/docs/9781464803352.

per year (USAID 2014). This supports drinking water supply, irrigation, fish farming, and other economic activities in the area.

Figure 2.5 shows consumption by different water-using sectors, excluding consumption of recycled water or reuse of waste and sewage water. Water consumption has fluctuated over time. Irrigation remains the largest consumptive user.

Figure 2.1 Spatial Distribution of Population and River Flow

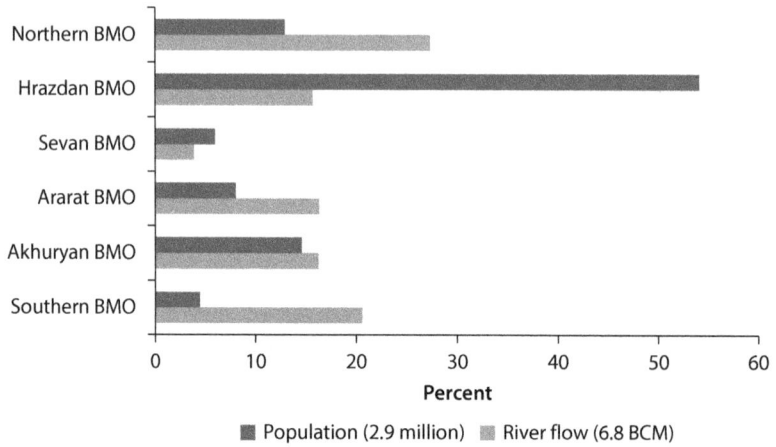

Source: USAID 2008b.
Note: BCM = billion cubic meters.

Figure 2.2 Long-Term Average Monthly Discharge

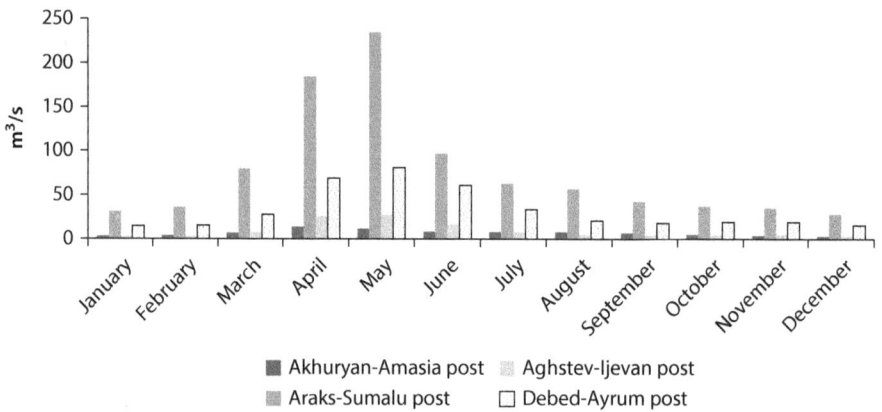

Source: Armenian State Hydrometeorological and Monitoring Service (ASHMS).

Figure 2.3 Time Series Monthly Discharge (Akhuryan-Akhurik station)

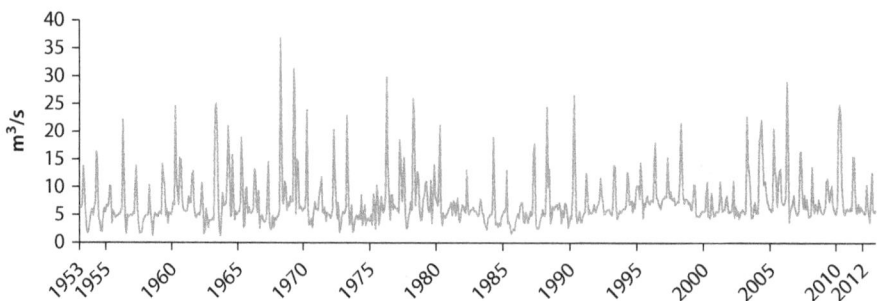

Source: ASHMS.

Figure 2.4 Water Abstraction by Source, 1995–2012

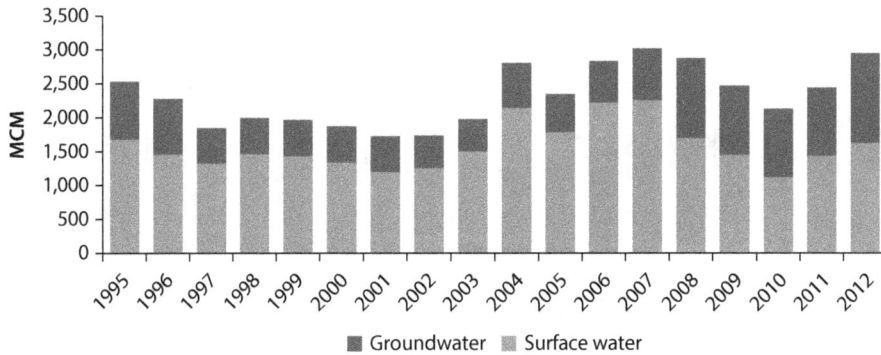

Source: National Statistical Service of Armenia.
Note: MCM = million cubic meters.

Table 2.2 Groundwater Resources of Armenia

| Basin | Area | Total groundwater resources | Of which: | | | | | |
| | | | Spring flow[a] | | Drainage flow[b] | | Deep flow[c] | |
	km²	MCM/yr	MCM/yr	Total (%)	MCM/yr	Total (%)	MCM/yr	Total (%)
Debed	3,790	506.4	113.3	22.4	356.2	70.4	36.9	7.3
Aghstev	1,730	192.8	44.0	22.8	85.9	44.5	62.9	32.6
Kura tributaries	477	54.0	19.7	36.5	29.2	54.1	5.1	9.4
Kasakh	1,480	426.5	129.1	30.3	68.2	16.0	229.2	53.8
Hrazdan	2,560	465.5	267.4	57.5	132.1	28.4	66.0	14.2
Lake Sevan	4,745	658.9	288.6	43.8	125.2	19.0	245.1	37.2
Azat	572	200.0	135.2	67.6	58.8	29.4	6.0	3.0
Vedi	633	39.1	15.0	38.4	14.7	37.6	9.4	24.0
Arpa	2,080	353.9	169.2	47.8	132.0	37.3	52.7	14.9
Akhuryan	2,784	367.1	142.8	38.9	85.9	23.4	138.4	37.7
Vorotan	2,030	544.0	171.9	31.6	251.9	46.3	120.2	22.1
Voghji	788	158.0	79.0	50.0	68.9	43.6	10.1	6.4
Meghriget	366	51.0	18.9	37.0	25.2	49.4	6.9	13.6
Total		**4,017.0**	**1,594.1**		**1,434.2**		**988.9**	

Source: Ministry of Nature Protection 2013, based on USAID 2008b; data are from the 1980s.
Note: MCM = million cubic meters.
a. *Spring flow* is artesian groundwater discharge. These values are based on field hydrogeological studies.
b. *Drainage flow* is base flow from shallow groundwater aquifers and is based on measurements in different river sections when there has been no precipitation.
c. *Deep flow* is calculated from the water balance.

Figure 2.5 Water Consumption by Sector

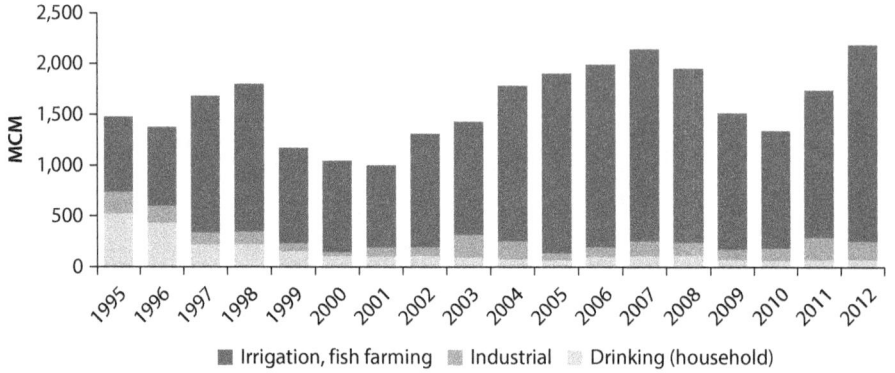

Irrigation, fish farming   Industrial   Drinking (household)

Source: National Statistical Service of Armenia.
Note: MCM = million cubic meters.

Irrigation and Drainage

Over recent decades, though the agriculture sector has added more value in absolute terms to the economy, its overall share of gross domestic product (GDP) has steadily decreased (around 18 percent in 2012) (figure 2.6). Yet, Armenia is still an agrarian society with the agriculture sector providing around 40 percent

Figure 2.6 Agriculture Value Added

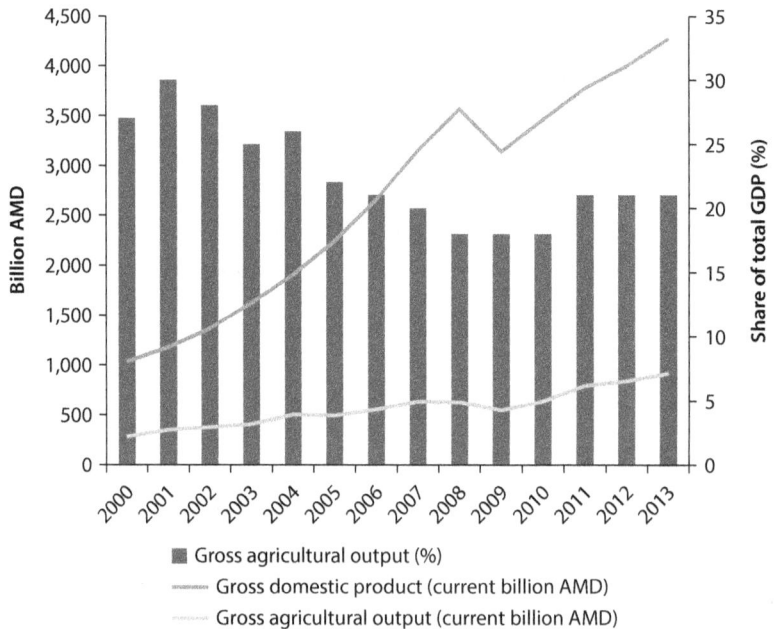

Gross agricultural output (%)
Gross domestic product (current billion AMD)
Gross agricultural output (current billion AMD)

Source: National Statistical Service of Armenia.

of total employment. Moreover, with important links to the growing food processing industry, agriculture will continue to play an important role in the Armenian economy.

Agriculture in Armenia is heavily dependent on irrigation. More than 80 percent of the gross crop output is produced on irrigated lands. Wheat, potatoes, and vegetables claim two-thirds of the total irrigated arable land. The consumption of irrigation water has fluctuated significantly over time, mainly due to fluctuations in overall water availability, and reached almost 2 billion cubic meters in 2012 (figure 2.7). Total irrigable area in Armenia is around 208,000 hectares. In 2005, the net income per hectare for wheat was 65,000 Armenian drams (US$156), twice as much as on rain-fed lands in the mountainous areas. Due to agroclimatic conditions, the most fertile regions are also the greatest consumers of irrigation water. At the same time, they show the lowest water productivity: while taking 80 percent of the country's irrigation water, they generate 53 percent of the Armenian gross crop output (figure 2.8) (World Bank 2013a).

Water user associations play an important role in agricultural water management. Currently, there are 42 water user associations (WUAs) responsible for about 195,000 hectares (out of a total of 208,000 hectares of irrigable lands in Armenia). In 2013, 130,524 hectares were actually irrigated under WUAs. This difference is primarily due to rain-fed areas, areas with poor intercommunity or intracommunity networks, and lack of cultivation. The operation of secondary and tertiary systems and small pumping stations and reservoirs has been transferred to WUAs. Two State water supply agencies (WSAs) operate the main large reservoirs, big pumping stations, and main canals,[3] and deliver bulk supplies to these WUAs. Since WUAs became operational, water supply has improved, the collection of water fees has increased, and there is an increasing conversion from low-value crops (e.g., wheat) to higher-value crops (e.g., fruits and vegetables). Table 2.3 summarizes the improvements over time, and map 2.2 and figure 2.9 show the areas irrigated by WUAs by location and by crop.

Figure 2.7 Water Consumption for Irrigation

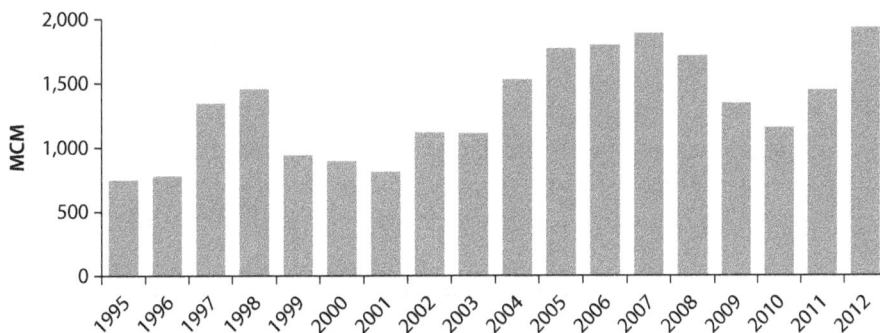

Source: National Statistical Service of Armenia.
Note: MCM = million cubic meters.

Figure 2.8 Irrigation Water Consumption and Agricultural Productivity by Province in 2010

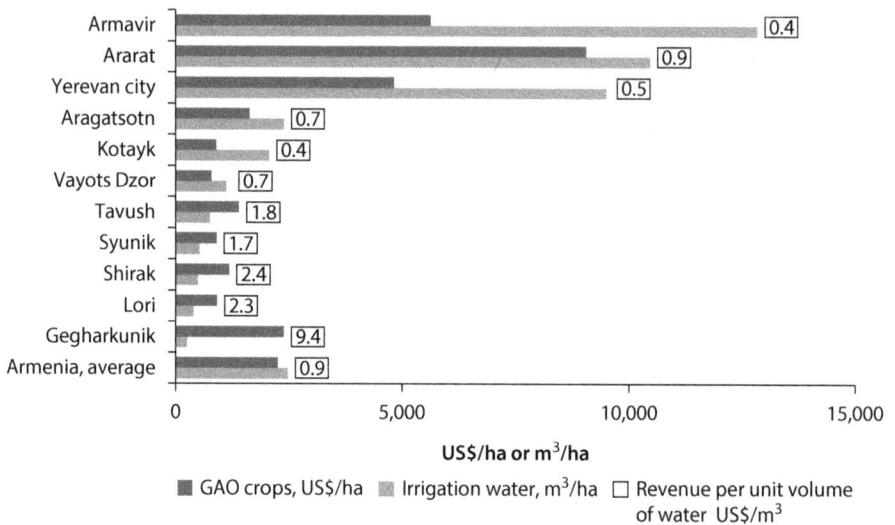

Source: Based on World Bank 2013a.
Note: GAO = gross agricultural output.

Table 2.3 Improvements after the Operationalization of Water User Associations, 2004–2013

	2004	2005	2006	2007	2008	2009	2010	2011	2012	2013
Irrigated area (ha)	113,366	125,648	123,298	125,632	128,860	128,076	129,194	129,406	130,180	130,524
Collection (billion AMD)	2.51	2.89	2.95	3.10	3.44	3.22	3.56	3.77	4.03	4.44
Collection rate (%)	56	66	69	73	68	87	82	83	78	86
High-value crops (%)	65	71	74	78	79	79	80	84	87	88

Source: Project implementation unit data.
Note: AMD = Armenian drams.

Water user associations are not yet financially sustainable and continue to depend on State subsidies. The irrigation service fee of WUAs is subject to a government-imposed ceiling. The current ceiling level is 11 Armenian drams per cubic meter of water, while the actual average cost is estimated at 17 drams.[4] The gap between the regulated fee and the actual cost is covered from the State budget. While the collection rate by WUAs averages 80 percent, actual cost recovery is estimated to be around 45 percent.[5] Current tariffs and subsidies do not encourage farmers to adopt more water- and energy-efficient practices or technologies. The water pricing system needs to be updated. Further financial strengthening of WUAs is a priority.

Agricultural water management is still subject to various inefficiencies. Most of the irrigation and drainage infrastructure built during Soviet times has not been adequately maintained. The budgets for rehabilitation and further infrastructure development decreased significantly from about 50 billion Armenian

Map 2.2 Irrigated Areas under Water User Associations, 2008

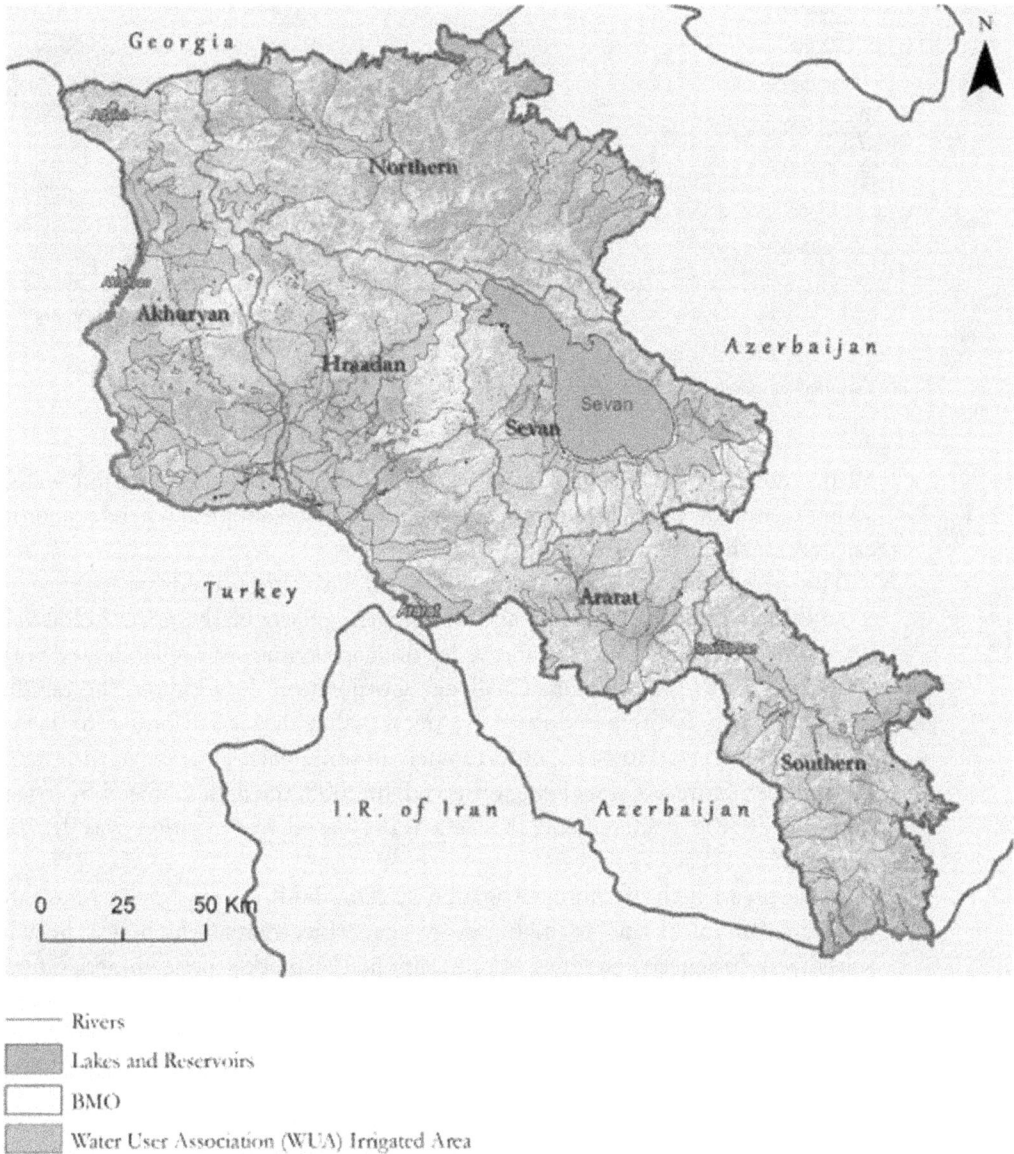

Rivers

Lakes and Reservoirs

BMO

Water User Association (WUA) Irrigated Area

Source: USAID 2008b.
Note: A full-color version of this map may be viewed at http://www.issuu.com/world.bank.publications/docs/9781464803352.

drams (US$120 million) per year during the Soviet era to 4 billion Armenian drams (US$10 million) per year on average in the period 1994–2011, including donor assistance. Operation and maintenance costs have been reduced from 25 billion Armenian drams (US$60 million) per year in the Soviet era to 8–10 billion Armenian drams (US$20–25 million) per year now (World Bank 2013a). As a consequence, water conveyance losses have gradually increased, to around

Figure 2.9 Water User Associations: Irrigated Area by Crop, 2012

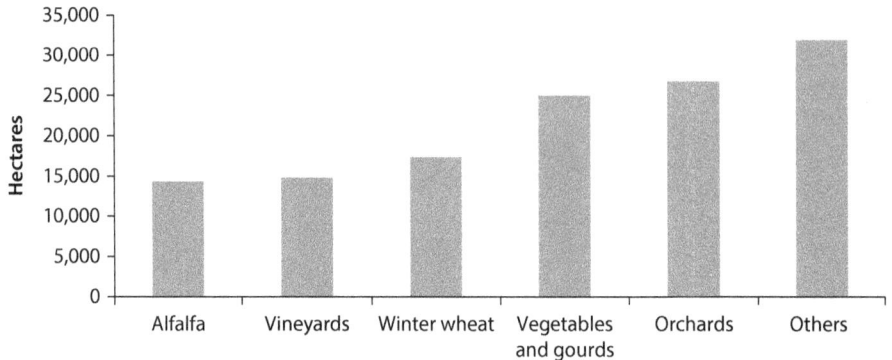

Source: Project implementation unit data.

59 percent in 2012.[6] Rehabilitation of irrigation canals is needed, and water-saving technologies, such as drip irrigation, need to be adopted where economically and technically justified.

The deterioration of the drainage system has also caused an increase in groundwater levels, salinization, and waterlogging, particularly in the Ararat valley. From 2005 to 2010, the Ararat valley drainage system was rehabilitated with support from the Millennium Challenge Corporation. In addition, the rapidly expanding fish farming industry in Ararat valley has contributed to lower groundwater levels. However, unfortunately in some places, excessive withdrawals from fish farms are now being observed. In 2006, the area salinized by irrigation was 20,400 hectares and the area waterlogged by irrigation was 18,700 hectares.[7]

Widespread high-lift pump irrigation systems built during Soviet times are now uneconomical due to high energy costs. Electricity, which was heavily subsidized during Soviet times, is now supplied at market price to agricultural water users. Pump irrigation systems are now being substituted with more energy-efficient gravity schemes. As a result, electricity spending by WSAs has decreased from 129 million kilowatt-hours to 25 million kilowatt-hours (84 percent reduction) (figure 2.10).[8]

Urban and Rural Water Supply

Domestic water consumption, which used to be the second-largest water user after irrigation, sharply decreased in the 1990s (figure 2.11). This dramatic drop is attributed to the introduction of water metering and a volumetric billing system. During Soviet times, domestic water bills were based on water pipe diameter and the number of household members. This practice was discontinued in 2000 when water meters were installed. The domestic water consumption data after 2000 better represent actual household water use. In 2012, domestic water consumption was 75.3 MCM per year,[9] or 25 cubic meters per capita per year.

Figure 2.10 Electricity Consumption for Irrigation

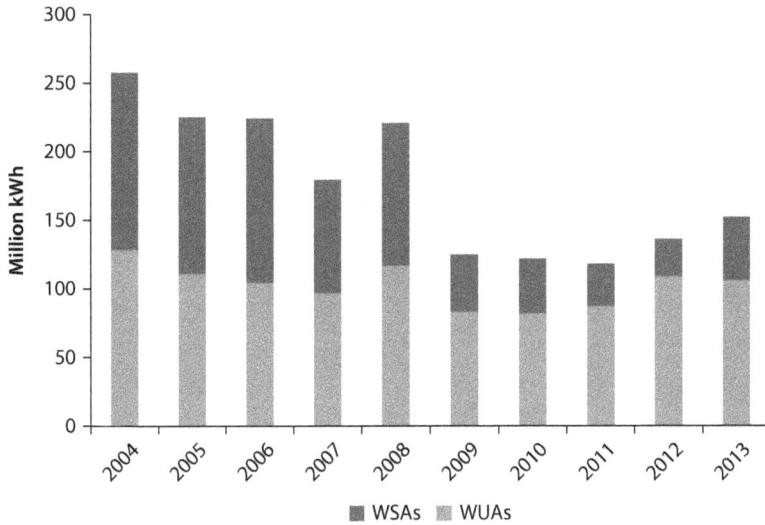

Source: Project implementation unit data.
Note: WSAs = water supply agencies; WUAs = water user associations; kWh = kilowatt-hours.

Figure 2.11 Water Consumption for Domestic Sector

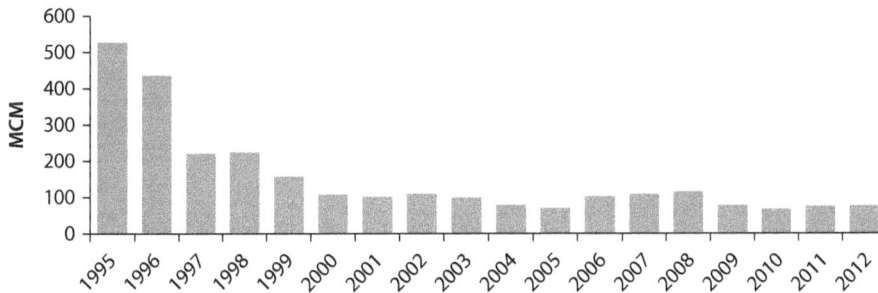

Source: National Statistical Service of Armenia.
Note: MCM = million cubic meters.

For many years after the collapse of the former Soviet Union, water supply and sanitation systems in Armenia were in a serious state of disrepair. The water supply system provided water only for a few hours a day. In the early 2000s, the government set rehabilitation of water supply infrastructure and achieving 24-hour water service as top priorities. Over the past decade, water supply in Armenia has greatly improved with the increased use of public-private partnerships. Currently, the majority of the population of Armenia is served by three water and wastewater utilities under public-private partnership arrangements (table 2.4). Outside those arrangements, 560 villages (about 500,000 people) have their own arrangements.

Table 2.4 Water Supply Utilities under Public-Private Partnership

	Water and sewerage company				
	Yerevan	Armenia	Shirak	Lori	Nor Akung
Ownership	Private company	State company	51% State shareholding and 49% municipal shareholding		
Management model	Centralized	Centralized	Decentralized (community involvement)		
Operator	Veolia, France	Saur, France	MVV consortium		
Contract mode	Management, lease	Management	Management		
Contract period	Until 2016	Until mid-2016	Until mid-2016		
Population served	1.17 million	0.91 million	0.36 million		
Loan	World Bank	World Bank	KfW		

Source: World Bank 2011b.
Note: KfW = KfW Development Bank.

Table 2.5 Performance Measures for Water Supply Utilities, before PPP versus 2009

Water and sewerage company	Water supply duration (hours)		Compliance with water quality requirements		Energy consumption (million kWh)		Collection efficiency (%)		Installed water meter (% of customers)		Unaccounted for water (%)	
	Before PPP	2009	Before PPP	2009	Before PPP	2009	Before PPP	2009	Before PPP	2009	Before PPP	2009
Yerevan	4–6	20.4	94.5	97.8	240.3	109.6	21	97.6	0.8	96	72	81.1
Armenia	4–6	12.8	93.8	98.4	64.4	46.6	48	84.1	40	72.3	74	83.6
Shirak	4.7	10.2	98.1	99.6	0.9	1.2	49	78	12	50	85	83
Lori	4	9.5	88	92	0.96	0.92	58	80	67	85	77	71
Nor Akung	4	22.3	100	100	7.2	4.0	47	97	20	93	87	70

Source: World Bank 2011b.
Note: PPP = public-private partnership; kWh = kilowatt-hours.

The public-private partnership approach has shown success, particularly with improving water supply duration, water meter installment, and collection efficiency. Compliance with water quality requirements has also improved and energy consumption has, in most cases, been reduced (table 2.5). However, levels of nonrevenue water have remained high (70–85 percent), of which approximately 45 percent is estimated to be technical losses, such as leakages due to the age and very poor state of the physical pipework and assets, and 40 percent comprises commercial losses, including nonpayment, underpayment, and theft (World Bank 2011b). Levels of nonrevenue water have not been taken as a main performance measure under the present public-private partnership contracts.

There remain some challenges that public-private partnerships alone cannot resolve. Although the collection rate is high, the tariff is still currently too low to provide sufficient funding to cover even the routine operation and maintenance costs and investment costs. The current tariff is 200 Armenian drams per cubic meter of water, which is considered low compared to regional and international norms of around 400 drams per cubic meter.[10] The deficit is covered by government subsidies: for instance, the Armenia Water and Sewerage Company received a subsidy of 8 million drams (US$19,000) per year from 2009 to 2011 for cost recovery (OECD 2012).

Moreover, while water supply has greatly progressed, sanitation has fallen behind. Wastewater collection and treatment systems are not sufficiently provided and operational, and wastewater is often discharged directly to water bodies or land, causing unhygienic conditions and water quality issues. Currently, 68 percent of the population (2 million, mostly urban) is connected to the sewerage network. There are 20 wastewater treatment plants, all built before the 1990s and inadequately maintained—either not operational or partially operational with mechanical treatment only. There is a need for major investment to rehabilitate and modernize wastewater treatment facilities and expand their coverage to rural areas (ADB 2011; World Bank 2011b).

Environment

Lake Sevan has environmental, economic, and social significance and is an important multipurpose water reservoir for irrigation, hydropower, and recreational uses. Lake Sevan, located in the central part of Armenia, is the largest lake in Armenia (almost 35 billion cubic meters) and one of the largest high-altitude lakes in the world. The lake is fed by 28 rivers and streams and is drained by the Hrazdan River. The lake outflow has been artificially regulated for irrigation and the Sevan-Hrazdan hydropower cascade since the 1930s. The level of Lake Sevan fell dramatically due to excessive use during the period from 1930 to the 1980s, resulting in serious environmental and ecological problems, including deterioration of water quality, destruction of natural habitats, and loss of biodiversity. A comparison between 2001 (the minimum level) and natural conditions in the 1930s shows a decrease in level by over 19 meters (from 1,915.65 meters to 1,896.55 meters above the level of the Baltic Sea), a decrease in volume from 58.5 billion to 32.9 billion cubic meters (44 percent), and a reduction of the surface area from 1,416 to 1,236 square kilometers (13 percent) (UNECE 2003, chapter 2: Lake Sevan).

Starting in the 1980s, programs to stabilize and raise the lake level were initiated, including the use of the Arpa-Sevan tunnel to transfer up to 250 MCM per year from the Arpa River. In the period 2001–13, on average 152 MCM per year were transferred to Lake Sevan through the Arpa-Sevan tunnel. The government adopted two laws[11] in 2001 that recognized the importance of Lake Sevan and aimed to raise the level by 6 meters[12] by 2030. This would add an additional 8.8 billion cubic meters to the lake. In addition to the Arpa-Sevan tunnel, the Vorotan-Arpa tunnel was built to increase the inflow to the lake. The tunnel was commissioned in 2004, and has the capacity to transfer up to 165 MCM per year from the Vorotan River.[13] Moreover, the lake outflow has been limited to 170 MCM per year for irrigation purposes (figure 2.12).[14] The Sevan-Hrazdan hydropower plants operate on a seasonal basis only during the release of Lake Sevan water for irrigation purposes.

Figure 2.12 Water Releases from Lake Sevan and Lake Level

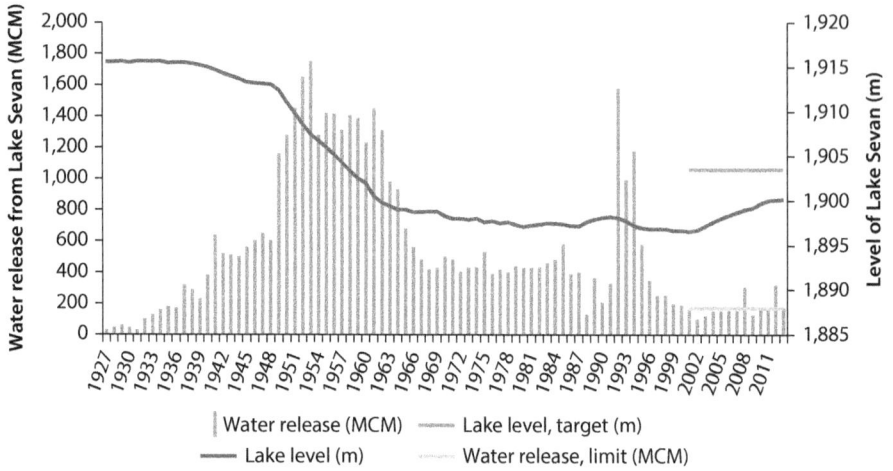

Source: State Committee on Water Economy.
Note: MCM = million cubic meters.

As a result of these measures, the level of Lake Sevan has been steadily rising since 2001 (figure 2.12). Between 2001 and 2013, the lake level rose by 3.9 meters and its volume increased by 5.5 billion cubic meters. Various environmental indicators have also improved. In 2008, the Presidential Commission on Lake Sevan Issues was formed. However, due to continued overfishing, the lake's whitefish population has continued to decrease to near-extinction level. It was estimated at 30,000 tonnes in the early 1980s, 3,500 tonnes in early the 2000s, and only 8 tonnes in 2011 (box 2.1).

There are growing concerns with respect to the declining quality of water in the country. One main driver for this is the discharge of untreated or insufficiently treated wastewater into surface water bodies. From 2008 to 2012, the total wastewater volume doubled (from 375 million to 813 MCM per year), and untreated discharge increased seven times (from 42 million to 307 MCM per year) (figure 2.13).[16] Some of this increase can be attributed to improved measurement and the increase in discharge from fish farming. All wastewater treatment plants were built during Soviet times and are now outdated, in need of rehabilitation, and are energy intensive and expensive to operate. Most plants have stopped operating and a few are applying mechanical treatment only. The growth of the mining industry has resulted in another potential source of pollution (for example, heavy metals) to water bodies.

Water–Energy Nexus

Water resources play a critical role in the energy sector. Armenia depends on power generation from thermal, hydro, and nuclear sources (figure 2.14). The total installed capacity is 3,603 megawatts, including 1,756 megawatts of

Box 2.1 Fisheries in Lake Sevan

The fish species endemic to Lake Sevan are Sevan trout (*Salmo ichchan*), Sevan khramulya (*Varicorhinus capoeta sevangi*), and Sevan barbel (*Barbus lacerta goktchaicus*). During the Soviet period, common whitefish (*Coregonus lavaretu*), crucian carp, and crayfish were introduced to the lake in order to increase fish catches. However, due to years of unrestrained fishing, the fish stock has drastically decreased. According to an assessment by the Institute of Hydroecology and Ichthyology, the fish reserves in Lake Sevan decreased from 30,000 tonnes in the early 1990s to 7–8 tonnes in 2012 (EcoLur 2012). Different types of trout have dramatically reduced. All three endemic species are listed in the Red Book of Armenia (FAO 2011). In addition to the decreasing stock, the size of captured fish has also decreased. The average weight of whitefish in 1997 was 222 grams, compared to 904 grams in the 1970s (FAO 2011).

There have been different measures taken to reverse these trends in Lake Sevan. The Sevan National Park was established in 1978 and the area was designated a Ramsar site in 1993.[15] A ban on fishing (starting in 2002), particularly for trout and whitefish, is routinely applied for the winter months or for a year-long period. A plan to construct a fish hatchery for Sevan trout production is being discussed. Finally, though fishing is prohibited, enforcement is weak and economic alternatives for local fishers are not available (photo B2.1.1).

Photo B2.1.1 Fish Selling Stall, Lake Sevan

Source: ©World Bank/Ju Young Lee. Used with Permission; further permission required for reuse.

Figure 2.13 Wastewater Discharge

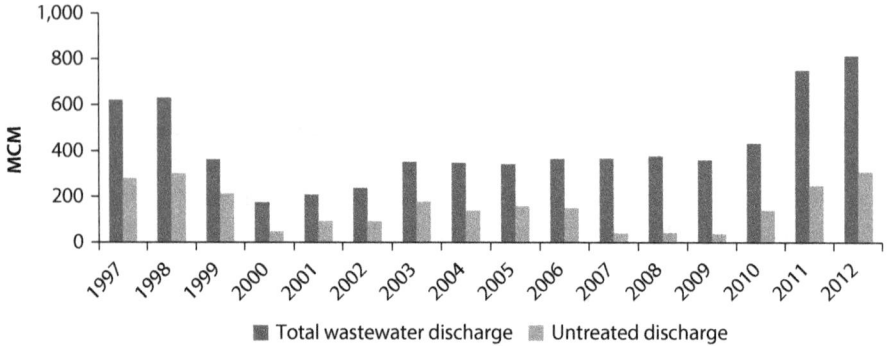

Total wastewater discharge   Untreated discharge

Source: National Statistical Service of Armenia.
Note: MCM = million cubic meters.

Figure 2.14 Electricity Production by Type

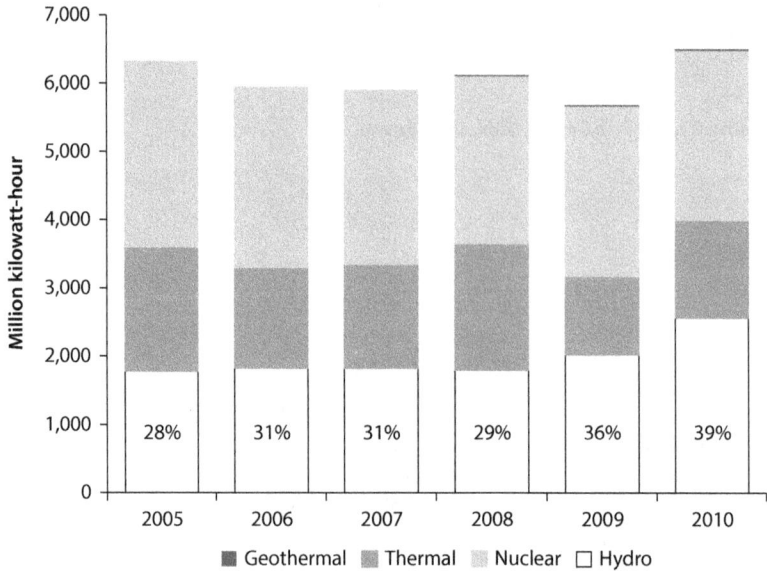

Geothermal   Thermal   Nuclear   Hydro

Source: United Nations 2010.

thermal power, 1,032 megawatts of hydropower, and 815 megawatts of nuclear power (figure 2.15) (World Bank 2011a). Armenia has great potential for hydropower from its mountains and fast-flowing rivers. Recent analysis finds that an additional 250–300 megawatts of generation is possible from small hydropower plants (Danish Energy Management 2011).

There are two large cascades and a number of small hydropower plants (table 2.6). Since the adoption of the Law on Privatization of State Property in 1997, all hydropower systems have been gradually privatized (especially small

Figure 2.15 Net Installed Capacity

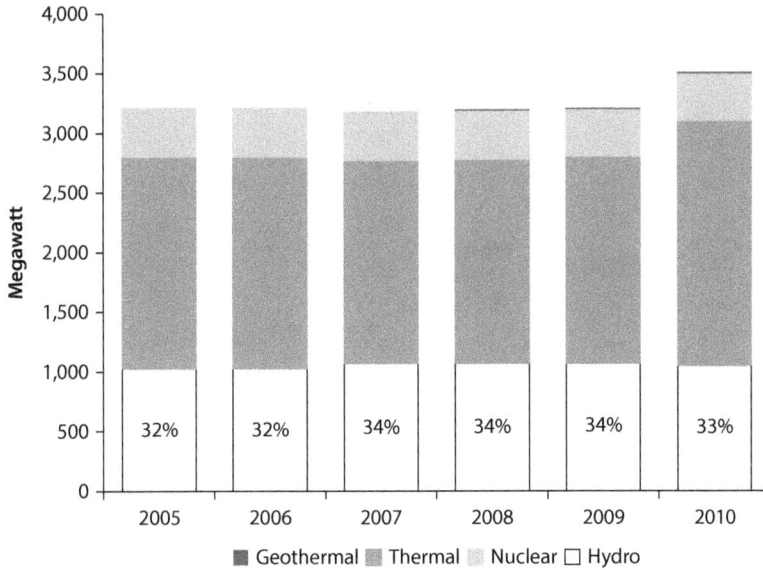

Source: United Nations 2010.

Table 2.6 Hydropower in Armenia

	Installed capacity (MW)*	Actual power generation (GWh) 2012†	Commissioning date*	Ownership*
Sevan-Hrazdan cascade	561	632.3	1940–62	RAO Nordic
Vorotan cascade	400	1,118.8	1970–89	Contour Global Hydro Cascade
Small hydropower plants	263.26‡	574.7a	–	Private owners
Total	**1,224.26**	**2325.8**	–	–

Sources: World Bank 2011a (*); Arthur Kochnakyan, World Bank (†); Public Services Regulatory Commission of Armenia (www.psrc.am) (‡).
Note: MW = megawatts; GWh = gigawatt-hours.
a. includes 62.1 GWh from Drozaget hydropower plant.

hydropower plants). The amendment to the Energy Law in 2001 provided for guaranteed 15-year power purchase agreements. In 2004, the Public Services Regulatory Commission adopted a feed-in tariff to drive forward investment in small hydropower plants, especially for run-of-the-river types. The Water Code was also amended to extend the water permit period to 5–10 years for small hydropower plants. As a result, the last decade has witnessed a major growth in the numbers of private small hydropower plants, spread throughout the country (map 2.3). As of 2012, there are 129 existing small hydropower plants with a capacity of 210 megawatts, and 75 more under construction with a capacity of 156 megawatts (for a total of 366 megawatts).[17] Currently, small hydropower plants provide about 6 percent of the total electricity in Armenia.

Map 2.3 Distribution of Small Hydropower Plants

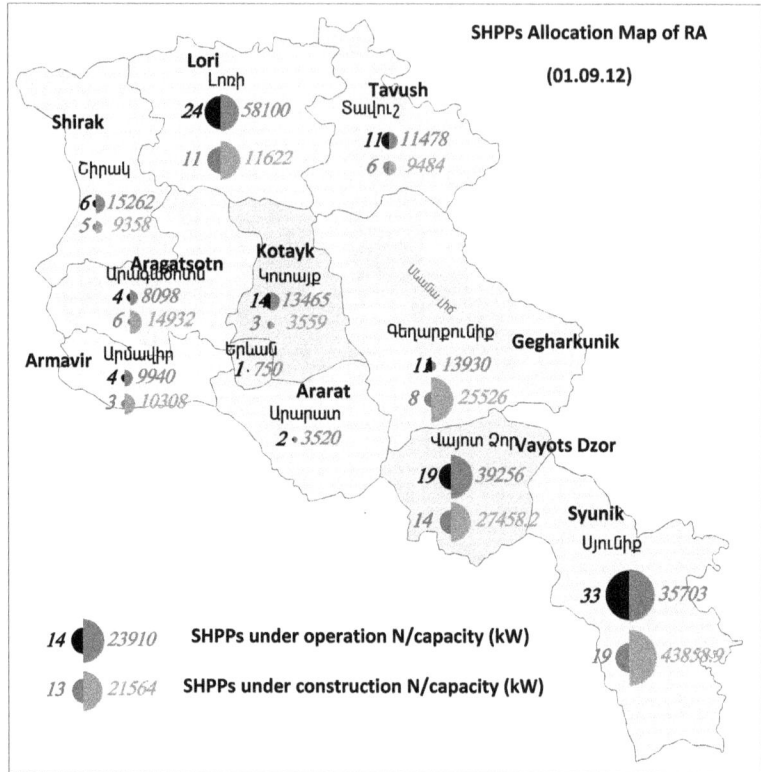

Source: Hydroenergetica.
Note: kW = kilowatts. A full-color version of this map may be viewed at http://www.issuu.com/world.bank
.publications/docs/9781464803352.

Some have raised concerns regarding the impact of existing and future small hydropower plants on water resources and environmental sustainability. The current permit system for small hydropower plants requires an environmental impact assessment and a study of streamflow limitations to satisfy minimum environmental flow requirements and other existing water demands. However, the environmental impact assessment is only partial and does not consider the basinwide cumulative impact. The procedures for the calculation of minimum environmental flow[18] may not be adequate, as they do not take into account seasonality and the site-specific ecosystem requirements. Moreover, small hydropower plants are not well monitored in relation to the water use permit system. Further analysis is needed on this issue.

Other power plants—thermal and nuclear—also use water resources for cooling purposes. In 2012, cooling water withdrawal and consumption were estimated to be 4–7 MCM and 3.3–6.4 MCM per year, respectively, for thermal power plants, and to be to be 23 MCM and 13 MCM per year, respectively, for nuclear power plants (table 2.7).

Table 2.7 Cooling Water Withdrawal and Consumption Estimates

	Cooling water requirement per unit power generation for recirculating system			Cooling water estimate for Armenia	
	Withdrawal (gallon/MWh)	Consumption (gallon/MWh)	Power generation 2012 (MWh)	Withdrawal (MCM)	Consumption (MCM)
Thermal power plant (natural gas steam turbine)	950–1,460[a]	662–1,170[a]	797,200	3–4	2.0–3.5
Thermal power plant (natural gas combined cycle)	150–283[a]	130–300[a]	2,577,028	1–3	1.3–2.9
Nuclear power plant	2,659[b]	1,481[b]	2,310,900	23	13
Total				**27–30**	**16.3–19.4**

Note: MWh = megawatt-hours; MCM = million cubic meters.
Sources:
a. Union of Concerned Scientists 2013.
b. USAID 2008a (calculated from 2007 data).

Climate Change

Compared to other countries in the region, Armenia is highly vulnerable to climate change. According to the World Bank (2013b), Armenia shows high exposure, high sensitivity, and limited adaptive capacity to climate change. Studies show that climate change is already occurring in Armenia. The Ministry of Nature Protection (2009) finds that temperatures have increased by 0.85°C and precipitation has decreased by 6 percent in Armenia over the past 80 years. These changes in temperature and precipitation vary by region and season. Summer temperatures increased by 1°C during the period 1935–2007, whereas winter temperatures increased by 0.04°C. The Ararat valley region has become more arid, while the southern and northwestern areas and the Lake Sevan basin have had a significant increase in precipitation during the last 70 years. The frequency of severe hydrometeorological phenomena (here defined as frosts, hailstorms, heavy rainfall events, and strong winds) also increased by 1.2 cases per year (statistically significant) over the period 1975–2005 (figure 2.16) (Ministry of Nature Protection 2009).

Future climate projections indicate continued increases in temperature and decreases in precipitation (table 2.8). The Ministry of Nature Protection (2010) projects a 4°C increase in temperature and 9 percent reduction in precipitation by 2100. The Ararat valley region is projected to experience higher warming than the rest of the country for all seasons. Temperature increases are predicted to be highest in the summer, and precipitation decline to be the greatest in the summer, the key agricultural season. The largest changes in precipitation are expected at altitudes higher than 1,700 meters, which are the main areas of river flow formation (Ministry of Nature Protection 2010). On average (across different models; see appendix C for projected precipitation and temperature changes by 2050 across the range of global circulation models), overall water resources availability is expected to reduce (ENVSEC and UNDP 2011). Increased air

Figure 2.16 Extreme Hydrometeorological Events in Armenia, 1975–2006

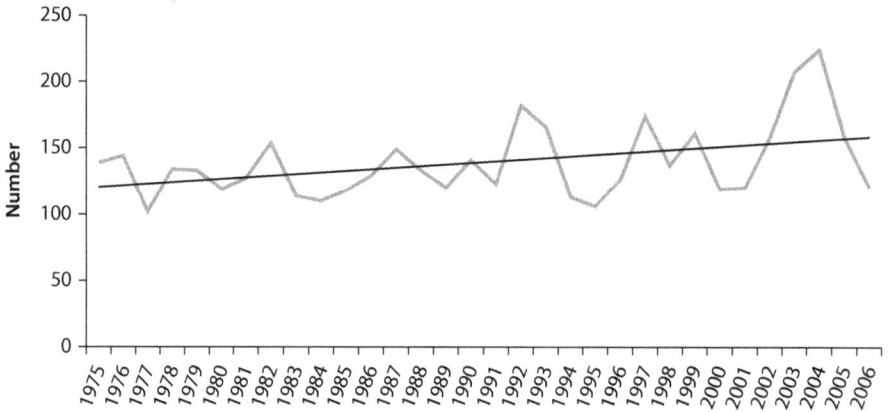

Source: Ministry of Nature Protection 2010.

Table 2.8 Climate Change Scenarios by 2100

Category	2030	2070	2100
Temperature[a]	+1°C	+2°C	+4°C
Precipitation[a] (%)	−3	−6	−9
Evaporation (%) (compared to 1991–2006)	+1.6	+2.5	+3.7
Snow cover (%) (compared to 1961–1990)	−7~11	−16~20	−20~40
River flow (%) (compared to 1961–1990)	−6.7 (or 0.3 BCM)	−14.5 (or 0.7 BCM)	−24.4 (or 1.2 BCM)

Source: Ministry of Nature Protection 2010.
Note: BCM = billion cubic meters.
a. 1935–2007 data were compared with respect to the base period 1961–90.

temperature and lower precipitation will increase evaporation rates and reduce winter snowpack and spring runoff, resulting in less river flow. Snow cover is expected to decrease as much as 20–40 percent by the end of the 21st century and river flow by almost a quarter. However, there are some regional differences; in some basins, such as the Vorotan and Voghji, river flow may increase (Ministry of Nature Protection 2009).

The impacts of climate change will be particularly severe for Lake Sevan. The 28 rivers and streams that flow into the lake are expected to decrease by 41 percent or 310 MCM by 2100 (table 2.9). The tunnels that divert water to Lake Sevan may also face reduced flows at the source (for example, Arpa River flow is projected to decrease by 66 percent by 2100). Furthermore, due to the reduction in volume and increase in air temperatures, water quality may deteriorate (Ministry of Nature Protection 2010).

In the agriculture sector, the most climate-sensitive sector, crop yields are predicted to decline and irrigation demands to increase with climate change. The

Table 2.9 Predicted Changes in the Main Elements of Lake Sevan Water Balance

Date	Precipitation			Evaporation			River flow		
	MCM	MCM change from the baseline	Change (%)	MCM	MCM change from the baseline	Change (%)	MCM	MCM change from the baseline	Change (%)
1961–1990 (baseline)	457	n.a.	n.a.	1,076	n.a.	n.a.	758	n.a.	n.a.
2030	449	−8.0	−1.8	1,158	82.0	7.6	665	−93.0	−12.3
2070	445	−12.0	−2.6	1,192	116.0	10.8	559	−199.0	−26.3
2100	436	−21.0	−4.6	1,268	192.0	17.8	449	−309.0	−40.8

Source: Ministry of Nature Protection 2010.
Note: MCM = million cubic meters. n.a. = not applicable.

Ministry of Nature Protection (2010) estimates that by 2030, yields of the main agricultural crops will decrease by 8–14 percent without adaptation (9–13 percent for cereals, 7–14 percent for vegetables, 8–10 percent for potatoes, and 5–8 percent for fruits). In order to maintain crop yields, substantially more irrigation will be needed. For example, in the Ararat valley region, irrigation water requirements for vegetables are predicted to increase by 38–42 percent by 2100 (UNDP 2011). However, with overall water resources availability expected to decline, these demands may be difficult to fully meet in the future. According to the Ministry of Nature Protection (2009), a 25 percent reduction in river flow is projected to result in a 15–34 percent reduction in the productivity of irrigated cropland (average 24 percent). The total future losses to the agricultural sector are estimated at around 75 billion to 170 billion Armenian drams (US$180 million to US$405 million). This is equivalent to a loss of 2–5 percent of GDP (in 2009), or more if indirect losses (for example, food processing industry, input markets) are also included.

The energy sector will also be affected, as Armenia uses its rivers for hydropower generation and cooling water for nuclear and thermal power plants. In particular, the country's energy program to further develop hydropower and increase the energy dependency on hydropower could be at risk. Reduced river flows both in time and space coupled with an increased demand for irrigation water should be taken into account in future hydropower planning.

Climate change is also likely to decrease water supply in transboundary basins. Future streamflow is assessed to decrease by 45–65 percent in the Khrami-Debed basin (Armenia/Georgia) and by 59–72 percent in the Aghstev basin (Armenia/Azerbaijan) by the end of the century (UNDP 2011).

Notes

1. Usable water resources of 9 billion cubic meters per year (USAID 2008b) divided by 2012 population of 2.9 million.

2. FAO AQUASTAT database: http://www.fao.org/nr/water/aquastat/main/index.stm.

3. The current inventory of irrigation infrastructure includes 3,000 kilometers of primary and secondary canals, 18,000 kilometers of on-farm or tertiary canals, 400 pumping stations, and 2,200 deep and shallow wells (World Bank 2013a).

4. The actual cost of water varies significantly from one scheme to another, subject to climatic, agronomic, and topographic conditions. In some cases, it may go up to 30 drams per cubic meter when more pumping is needed (World Bank 2013a).

5. It is important to differentiate between the cost recovery of the whole irrigation system, which currently is estimated at the level of 45 percent (in 2011 the overall operation and maintenance expenses of the system, including water supply agencies, were 8.5 billion drams or US$ 20.5 million, and the amount collected by water user associations was 3.85 billion drams or US$ 9.3 million), and the collection rate, which on average is 80 percent. This means that if the ceiling is removed or increased the cost recovery may improve significantly, as in general water users pay for the received services (World Bank 2013a).

6. Project implementation unit data.

7. FAO AQUASTAT database: http://www.fao.org/nr/water/aquastat/main/index.stm.

8. Project Implementation Unit/Tigran Ishkhanyan.

9. National Statistical Service of Armenia.

10. The current tariff level needs to increase by 33 percent in 2014 to achieve full cost recovery for operation and maintenance, debt service, and depreciation by 2022 (World Bank 2011b).

11. On Lake Sevan (May 15, 2001) and on Adoption of the Annual and Complex Programs of Activities for the Use, Protection, Reconstruction, and Reproduction of the Lake Sevan Ecosystem (December 14, 2001).

12. To 1,903.5 meters above the level of the Baltic Sea, the minimum level required to improve lake conditions, according to the calculation of local scientists.

13. The Vorotan-Arpa tunnel transfers water from the Vorotan River to the Kechut reservoir, which then releases water to Lake Sevan through the Arpa-Sevan tunnel.

14. An exception can be made at the request of the Ministry of Agriculture and with parliamentary approval, for example in a drought year. In 2014, it is approved to abstract up to 240 million cubic meters.

15. Wetland of international importance as designated by the Convention on Wetlands of International Importance Especially as Waterfowl Habitat (Ramsar Convention).

16. National Statistical Service of Armenia.

17. Personal communication with Inessa Gabayan, director at Hydroenergetica.

18. A new method of estimating minimum environmental flow (consecutive 10-day minimum flow from historical data) was introduced in 2011 to replace the old method (75 percent of the 95th percentile of previously recorded monthly water flow levels).

References

ADB (Asian Development Bank). 2011. *Armenia Water Supply and Sanitation: Challenges, Achievements, and Future Directions.*

Danish Energy Management. 2011. *Renewable Energy Roadmap for Armenia: Task 4 Report.* Prepared for Armenia Renewable Resources and Energy Efficiency Fund (R2E2), May 2011.

EcoLur. 2012. "Ban on Fishing in Sevan Exists, But Already No Fish." *EcoLur, New Information Policy in Ecology*, August 24. http://www.ecolur.org/en/news/sevan /ban-on-fishing-in-sevan-exists-but-already-no-fish/4104/.

Falkenmark, M. 1989. "The Massive Water Scarcity Threatening Africa: Why Isn't It Being Addressed?" *Ambio* 18 (2): 112–18.

FAO (Food and Agriculture Organization of the United Nations). 2011. *Review of Fisheries and Aquaculture Development Potentials in Armenia.*

Ministry of Nature Protection. 2009. *Vulnerability of Water Resources in the Republic of Armenia under Climate Change.* Prepared under the United Nations Development Programme/Global Environment Facility (UNDP/GEF) project implemented by UNDP Armenia and executed by the Ministry of Nature Protection.

———. 2010. *Second National Communication on Climate Change.* Report prepared for United Nations Framework Convention on Climate Change. Yerevan: Government of Armenia, Ministry of Nature Protection.

———. 2013. *Feasibility of the Master Plan for Integrated Water Resources Management in the Six Water Basin Management Areas of Armenia.*

National Statistical Service (multiple years). *Statistical Yearbook of Armenia: Natural Resources and Environment.* Yerevan: Government of Armenia, National Statistical Service. http://www.armstat.am/en/?nid=45.

OECD (Organisation for Economic Co-operation and Development). 2012. *Economic Instruments for Water Management in the Debed River Basin: Issues and Options.*

Ueda, S. 2012. *Armenia Water Resources and Dam Sector Mission Report.* Back-to-Office Report, World Bank.

UNECE (United Nations Economic Commission for Europe). 2003. *National Report on the State of the Environment in Armenia in 2002.*

Union of Concerned Scientists. 2013. "How It Works: Water for Power Plant Cooling." http://www.ucsusa.org/clean_energy/our-energy-choices/energy-and-water-use /water-energy-electricity-cooling-power-plant.html#ftn1.

United Nations. 2010. *Electricity Profiles (2005–10) for Armenia.* United Nations Statistics Division.

USAID (United States Agency for International Development). 2008a. *Armenia New Nuclear Unit Environmental Background Information Document.*

———. 2008b. *Water Resources Atlas of Armenia.*

———. 2014. *Assessment Study of Groundwater Resources of the Ararat Valley.* Final report, prepared under USAID Clean Energy and Water Program.

ENVSEC (Environment and Secutiry Initiative) and UNDP (United Nations Development Programme). 2011. *Regional Climate Change Impacts Study for the South Caucasus Region.*

World Bank. 2011a. *Energy Sector Note for Republic of Armenia.* Washington, DC.

———. 2011b. *Water Sector Note for Republic of Armenia.* Report 61317-AM, Washington, DC.

———. 2013a. *Agriculture and Rural Development Policy Note for Armenia.* Report AUS1680, Washington, DC.

———. 2013b. *Building Resilience to Climate Change for the South Caucasus Region Agriculture Sector,* Washington, DC.

CHAPTER 3

A Decade of IWRM Reform

Key Challenges

- A major challenge will be the decentralization and institutional strengthening of nascent institutions (for example, the WRMA, BMOs, and water user associations).
- Improved implementation and administration of the water use permit system is needed.
- Clarification is needed on the evolving roles and responsibilities of the WRMA and BMOs.
- Development of river basin management plans will be a critical IWRM tool.
- Harmonization of different agencies responsible for surface water and groundwater quantity and quality monitoring is needed.
- There are gaps in laws that have not yet been enacted but are authorized (or are inadequately authorized) by the Water Code.
- Both technical and financial barriers exist to the establishment of the programmatic systems necessary to the attainment of the Water Code's stated purposes.

Legal and Policy Basis for Water Resources Management

Over the last 10 years, Armenia has achieved significant legislative and institutional reforms in terms of water resources management and protection. Notable among these are the adoption of the updated Water Code in 2002, the Law on Water User Associations and Federations of Water User Associations in 2002, the Law on the Fundamental Provisions of the National Water Policy in 2005, and the Law on the National Water Program in 2006. These measures establish the principles and mechanisms needed to implement integrated water resources management (IWRM) in the country. In general, these laws are quite extensive and comprehensive in scope and serve as a strong foundation for planning and management in the water sector. It is worth highlighting the stated purpose and objectives of the 2002 Water Code (box 3.1).

Box 3.1 Water Code: Stated Purpose and Objectives

The main purpose of this Code is the conservation of the national water reserve, the satisfaction of water needs of citizens and economy through effective management of usable water resources, securing ecological sustainability of the environment, as well as the provision of a legal basis to achieve the objectives of this Code. The objectives of this Code are:

• Establishment of appropriate water resources management mechanisms;
• Conservation and protection of water resources, including mitigation of pollution, maintenance and supervision of water standards and water level of the national water reserve;
• Prevention of water's harmful impact;
• Ensuring water resources assessment;
• Ensuring water supply to population and economy in necessary quantity and quality by regulated tariffs;
• Safe and smooth working of water supply and wastewater systems, and provision of normal conditions for their use, maintenance, and supervision;
• Provision of conditions for hydrotechnical structures, and safe and smooth use, maintenance, and supervision;
• Organization of management, protection, and development of water systems.

The Code consists of 17 chapters and 121 articles. It is extensive in scope and establishes the basic principles of management, use, and protection of water resources and water systems, including (a) satisfaction of the basic vital needs of present and future generations; (b) maintenance and increase of the volume of the national water reserve;[1] (c) protection of aquatic and related ecosystems and their biological diversity, and recognition of the integrated and interconnected relationship between land, air, water, and biological diversity; and (d) regulation of water use through water use permits. The Code establishes various water resources management institutions, including the National Water Council (NWC) and its Dispute Resolution Council, the Water Resources Management and Protection Body, the Water Systems Management Body, and the Regulatory Commission. It also makes provision for a water policy (article 15) and water program (article 16), and establishes the need to develop basin management plans (article 17) and the monitoring system for the country (article 19), including the State Water Cadastre. Other provisions of the Code include the use and management of State-owned water systems (articles 48–62),[2] conditions of use and protection of transboundary water resources (articles 63–65), the use of and compliance with water quality standards (articles 66–70), economic incentives and the system of payment (articles 76–81), and emergency situations related to water-related disasters (articles 93–97).

Of particular interest is article 121, which presents a list of five laws and 66 bylaws to be established and put in place. Though this list is extensive in nature and does provide some degree of prioritization, the timely achievement of many of these has proven to be unrealistic due to the complexity of the activities and financial constraints. This includes, for example, the procedures for the use and presentation of water resources of international significance under special protection, procedures for the irrigation of agricultural lands with wastewater, and establishment of water basin management areas and approval of their management plans. Some of these activities are identified in the National Water Program.

Following the Water Code, in 2005 the Law on the Fundamental Provisions of the National Water Policy was adopted. This law provides greater definition and clarity on key aspects, including setting water resource use and protection priorities, establishing a broad procedure for demand estimation and water resources assessment, outlining additional water policy principles (not covered in the Water Code), and highlighting the centrality of the water basin management plan. Water allocation is clearly defined in this law in the following order: national water reserve (this is defined more clearly in the subsequent National Water Program); traditional (historical, nonextractive uses); water resource uses under current contractual arrangements; and domestic, agricultural, hydropower and energy generation, industrial, and recreational use. The law also establishes that water allocation among users should aim to maximize the total (economic, social, and environmental) value of the water resource. Additional water policy principles to guide water management include the use of good science, meeting basic needs, use of water pricing and economic instruments, integrated assessment (including environmental, cultural, social, and economic values), ecological balance of the environment, user pays and polluter pays principles, cost recovery, use of water quality norms, and transparency and public participation. Finally, this law establishes guidance on the development of a National Water Program of activities.

In 2006, the Law on the National Water Program was adopted. This law provides further clarity on various issues, including definition of the various types of "reserves" (table 3.1), classification of water systems and identification of those of State significance, assessment of water demand and supply, development of a strategy for storage, distribution, and use of water resources, delineation of the

Table 3.1 Water Resources and Reserves of Armenia

Basin management organization (BMO)	Usable water resources (MCM)	Strategic water reserve (MCM)	National water reserve (MCM)
Northern BMO	1,897	59.2	63.3
Hrazdan BMO	733	229.3	254.1
Sevan BMO	2,068	500.0	34,583.6
Ararat BMO	1,306	229.0	245.3
Akhuryan BMO	1,602	564.0	608.2
Southern BMO	1,443	90.5	101.1
Total	**9,049**	**1,672.0**	**35,855.6**

Source: USAID 2008b.
Note: BMO = basin management organization; MCM = million cubic meters.

Toward Integrated Water Resources Management in Armenia • http://dx.doi.org/10.1596/978-1-4648-0335-2

issues in various water subsectors (for example, water supply and wastewater collection, irrigation, hydropower), development of water standard norms, and improvement of water resources monitoring. Short-term (until 2010), medium-term (2010–15), and long-term (2015–21) measures for implementation of the National Water Program were also identified.

To date, many of the National Water Program measures have not been fully implemented or have been largely supported by international donors and through bilateral assistance (WRMA 2011). Thus, technical capacity and the necessary internal budgets to implement these measures have been insufficient. Short-term measures to be taken, which were to be completed by 2010, remain mostly incomplete (table 3.2). The Lake Sevan Action Plan is the only measure

Table 3.2 Implementation Status of Short-Term Measures of the National Water Program

Issues and short-term measures	Implementation status[a]		
	1	2	3
Legal requirements			
1. Intersectoral harmonization and improvement of the existing legislation			
2. Establishment of an interagency standing commission within the National Water Council (NWC) to discuss amendments to be made to the legal acts			
Institutional development			
3. Review and implementation of developed recommendations related to overlaps and gaps in the roles and responsibilities			
4. Adjustment and improvement of the mechanisms for interagency cooperation and coordination by the NWC			
5. Development of a program for institutional development of the basin management organizations (BMOs)			
Water resources management needs			
6. Development and testing of a pilot monitoring system in one basin management area			
7. Development of a monitoring strategy and a national program			
8. Reestablishment of the groundwater resources monitoring system in Armenia			
9. Improvement of the existing water use permit regulations, and establishment of criteria for priority of water use application			
10. Development of criteria and guidelines for environmental impact assessment as part of the water use permit application process			
11. Development and implementation of a short-term program for the State Water Cadastre			
12. Ensuring public awareness and participation in the planning and management of water resources at the national and basin levels			
13. Development and implementation of strategies for establishment of basin public councils, and technical capacity building			
14. Implementation and continuous monitoring and assessment of the National Water Program			
15. Establishment of a monitoring system for the program implementation			
16. Capacity building in the water resources management agency (WRMA) and basin management organizations for integrated water resources management (IWRM)			

table continues next page

Table 3.2 Implementation Status of Short-Term Measures of the National Water Program *(continued)*

Issues and short-term measures	Implementation status[a]		
	1	2	3
17. Development of a pilot RBMP and identification of information needs for one basin management area			
18. Review and improvement of the programs of measures for restoration, protection, reproduction, and use of the Lake Sevan ecosystem			
19. Clarification of up-to-date characteristics of water resources and water reserve components			
20. Adjustment and introduction of an international methodology for determination of norms for the limitation of impacts on water resources			
21. Development of a methodology for determination of aquatic ecosystem protection zones			
22. Development and implementation of programs for use of previously drained agricultural lands in the Ararat valley			
23. Implementation of works provided for under the program for reservoir construction			
24. Development of a strategy for water quality management			
25. Review and improvement of the existing approaches to spatial planning			
26. Development of a program for management of transboundary water resources			
Water systems management needs			
27. Study of water supply and wastewater collection services and implementation of programs to improve the provided services			
28. Development of programs aimed at enhancing the measures for the safety of hydrotechnical structures and reliability of operations			
29. Clarification of responsibilities for operation and protection of hydrotechnical structures of State significance			

a. Implementation status: 1 = Started; 2 = Progress; 3 = Completed.

that has been fully achieved. Moreover, pursuant to the requirements of the Water Code and the Law on the National Water Program, each year the State-authorized bodies responsible for implementation of the measures of the National Water Program should report to the government on progress in implementing these measures. This largely has not been done. An institutional mechanism for monitoring and assessing the National Water Program is not in place.

Progress has been slow with several critical measures to be implemented related to institutional and legislative support. This includes establishing clear mechanisms for interagency coordination, support to new institutions identified in the Water Code (e.g., WRMA, BMO, basin public councils), and establishing robust mechanisms for public participation in the planning and management of water resources. Also, though additional laws and over 120 regulations and bylaws have been issued since these initial reforms to provide further guidance and clarification on a number of matters, there still remain areas where harmonization and improvement to existing legislation is needed. Finally, progress has been slow with initiating some key knowledge generation measures. This includes the development of an overall monitoring and water quality strategy for the country, an up-to-date assessment of the overall water resources and water reserves situation, and the establishment of accepted criteria and guidelines for environmental impact assessment. From an operational perspective, specific measures to enhance the water permitting process and continuity with the State Water Cadastre are also lagging. Details of these short-term measures and the latest progress are given in appendix D.

Current Main Institutions for IWRM

Table 3.3 shows the main institutions for IWRM in Armenia.

The **National Water Council**, chaired by the prime minister of Armenia, is the highest advisory body for water resources management. It provides guidance on issues concerning the National Water Policy, National Water Program, and other legal aspects. Draft laws and amendments are submitted to this body. The potential power of this council is unique and singular. However, the National Water Council does not have direct staff (or a secretariat) to coordinate information, policy, and program recommendations. The **Dispute Resolution Council**, responsible for resolving disputes related to the issuance of water use permits, is under the National Water Council.

The **Ministry of Nature Protection** has overall responsibility for natural resources management and protection, including atmosphere, water, soil, flora and fauna, and forests. The **Water Resources Management Agency** (WRMA) under the Ministry of Nature Protection has the responsibility for implementing the government's water resources management and protection plans (for both surface water and groundwater) under the Water Code (2002). This includes providing water availability and use estimates, water use regulation and allocation, issuing water use permits, monitoring, developing river basin management plans (RBMPs), ensuring that environmental needs for water are being met, and classifying water bodies. The WRMA is also responsible for the maintenance of the State Water Cadastre. The WRMA (and suborganizations) requires the most technical and financial support to fulfill its mandate.

The WRMA has three divisions: (a) the Water Basin Planning Department, which participates in water resources protection planning and water distribution planning, develops medium-term water allocation plans, and links to communities

Table 3.3 Main Institutions for IWRM in Armenia

	Management and protection of water resources	Regulation of tariffs	Management of water systems
Authorized agency	WRMA	Public Services Regulatory Commission	State Committee on Water Systems
Main functions	Monitoring and allocation of water resources, strategic management and protection of water resources	Regulation of tariffs for noncompetitive water supply and discharge services in drinking, household, and irrigation water sectors, protection of consumers' rights	Management of water systems under State ownership; support to establishment of water user associations and unions of water users, arrangement of tenders on management of water systems
Enforcement tools	Water use permits	Water system use permits	Management contract

via its water basin management bodies; (b) the Water Cadastre Maintenance and Monitoring Division, which maintains information regarding water use permits; and (c) the Water Use Permitting Department, which manages the water use permit process.

In addition, under the WRMA, there are six **basin management organizations** (BMOs) responsible for interfacing between the WRMA and the local communities in the basins. The six BMOs are Sevan BMO, Hrazdan BMO, Northern BMO, Akhuryan BMO, Ararat BMO, and Southern BMO. Many of the BMO mandates are shared with other existing water resources management institutions, particularly the WRMA, in the areas of water use planning, permitting, compliance, and enforcement. The current interpretation is that the BMOs are subordinate to the WRMA and support the WRMA in administering its water protection and conservation responsibilities. Thus, BMOs are responsible for participating in development of water basin management plans, recording water use permits, ensuring water resources protection, assuring compliance with conditions set in water use permits, developing extraction regimes, and participating in the development of water allocation plans for their respective basin management areas. Further legislative clarity may be required as the capacity and role of BMOs evolves.

Also established under the National Water Program are **basin public councils,** which are meant to be advisory bodies to the BMOs and to provide an avenue for public participation. These, however, have no well-defined mission, structure, or procedures.

In 2010 the **Water Policy Division** was established under the Ministry of Nature Protection. It is in charge of formation of State policy on water resources protection, development of policy programs and strategic directions, and monitoring of their implementation. This includes serving as the lead agency to initiate the development of new laws and regulations as required by the Water Code.

The **State Committee on Water Economy**, under the Ministry of Territorial Administration, is another State body that is responsible more specifically for the management and operational use of State-owned water systems (for example, irrigation, water supply, and sanitation). There are also two closed joint stock companies (Sevan-Hrazdan and Akhuryan-Araks irrigation intake companies) that are responsible for the management and operation of irrigation systems in these areas, including the reservoirs, main canals, and major pumping stations. These companies have signed contracts with water user associations for supplying water.

Following the 2001 Law on Water User Associations and Federations of Water User Associations, and Resolution 314-N dated March 13, 2003,[3] State-owned irrigation systems and property were transferred to **water user associations**, of which there are currently 42 responsible for an irrigated area of around 195,000 hectares. The tasks envisioned for water user associations include operation and maintenance of the irrigation system and distribution of water among its members, water supply to member and nonmember water users located in the service area, implementation of construction works and restoration of water resources located in the service area, obtaining irrigation water from a water supplier (the **State Committee on Water Economy**) or the intake of water from natural water bodies, levying of fees from members and nonmembers for provided services, and procuring hydrotechnical equipment. These water user associations are still in the early stages of capacity development.

The **Public Services Regulatory Commission** implements tariff policy in the water sector. In particular, it issues water system use permits to noncompetitive water suppliers and defines the tariff to these users. Thus, the commission approves the retail tariffs for potable water supply, discharge, and wastewater treatment for the consumer services provided by the drinking water supply companies, as well as the tariffs for irrigation water supply to water user associations, federations of water user associations, and other users.

Other ministries with responsibility and a role (either direct or indirect) in water resources management include the Ministry of Agriculture, Ministry of Health Care (which includes the State Health Inspectorate[4] responsible for safeguarding the sanitary safety of the population in the drinking water sector), the Ministry of Emergency Situations (which includes the ASHMS), and the Ministry of Energy (which implements policy and strategies in the hydropower sector).

A system of tariffs and fees are used to regulate water uses (both consumptive and nonconsumptive). Tables 3.4 and 3.5 provide an overview of these current

Table 3.4 Water Tariffs and Fees by Sector

Economic instrument	Beneficiary	Management objective	Sector
Abstraction fee	WRMA	Ensuring rational use and efficient allocation of water resources, and maintaining minimum environmental flow	Drinking water (household), industrial, irrigation, fisheries sectors
Pollution fee	WRMA	Reducing pollution	Industry, urban wastewater supply, irrigation, fisheries sectors
Tariff	Private companies, local administrations, supply, and discharge companies	Ensuring sustainable water supply to population	Irrigation water supply, drinking water supply, and discharge
Fines and penalties	State Environmental Inspectorate (SEI)	Complying with water use permit conditions, pollution reduction, and ensuring minimum environmental flow	All entities holding water use permits

Source: Ministry of Nature Protection 2013.

Table 3.5 Water Abstraction Fees per Cubic Meter (in Armenian Drams)

	Surface water, excluding Lake Sevan	Surface water, from Lake Sevan	Groundwater, suitable for drinking	Groundwater, not suitable for drinking
Fisheries	1	1.5	1	1
Industrial	0.5	1.5	1	1
Drinking water (household)	0.5	1.5	1	1
Drinking water (supply companies)	0.025	1.5	0.05	1
Irrigation	0	0.2	1	0

Source: Ministry of Nature Protection 2013.

instruments and current rates (abstraction fees only), and table 3.6 shows the revenue collected and government institution expenditures. Current expenditures exceed what is collected through these various fees.

Challenges Ahead

With the institutional arrangements and supporting legislative environment largely in place (figure 3.1), the government of Armenia has a strong foundation for achieving IWRM. Looking ahead, however, there are areas where continued support both technically and financially is needed. According to OECD (2013), the water resources management, monitoring, and compliance assurance organizations (WRMA, BMOs, ASHMS, Environmental Impact Monitoring Center, Hydrogeological Monitoring Center, SEI) together receive annually around 500 million Armenian drams (US$1.2 million) for their water-related activities. It is estimated that about 1.7 billion drams (US$4.1 million) is required to fully and properly implement the tasks and responsibilities assigned to these agencies. Moreover, second-generation reforms are needed, which include support to the decentralization process, strengthening the water permit system, strengthening the monitoring system, and broad-based capacity building on IWRM, particularly with respect to river basin planning. Building the capacity of the WRMA will be critical in these regards. Moreover, since the inception of these reforms, some additional pressures have emerged highlighting the criticality of IWRM in the country. These include increasing concerns on transboundary rivers, emerging conflicts in the important Ararat valley, and government priorities on new storage infrastructure. These issues will be examined more closely in the next sections.

Table 3.6 Water Abstraction Fees, Fines, and Penalties Collected from 2010 to 2013

Items	2010 million AMD	2011 million AMD	2012 million AMD	2013 million AMD
Revenue collected				
Water abstraction fee	176.9	175.5	229.8	181.0
Water pollution fee	268.1	271.0	221.6	254.1
Administrative penalty	13.9	6.5	6.3	3.3
Damage compensation	9.2	3.7	15.6	4.7
Total revenue	**468.1**	**456.7**	**473.3**	**443.1**
Government institution expenditures				
WRMA	101.3	91.6	100.5	100.5
SEI (Water Division)	218.0	218.0	218.0	218.0
Environmental Impact Monitoring Center (surface water quality monitoring only)	42.3	61.5	65.6	66.3
Hydrogeological Monitoring Center	18.0	18.1	18.1	18.4
Hydromet service of the Ministry of Emergency Situations	108.1	108.9	109.0	113.1
Total expenditures	**487.7**	**498.1**	**511.2**	**516.3**

Source: Ministry of Nature Protection 2013.
Note: AMD = Armenian drams.

Figure 3.1 Institutional Framework on Water Management in Armenia

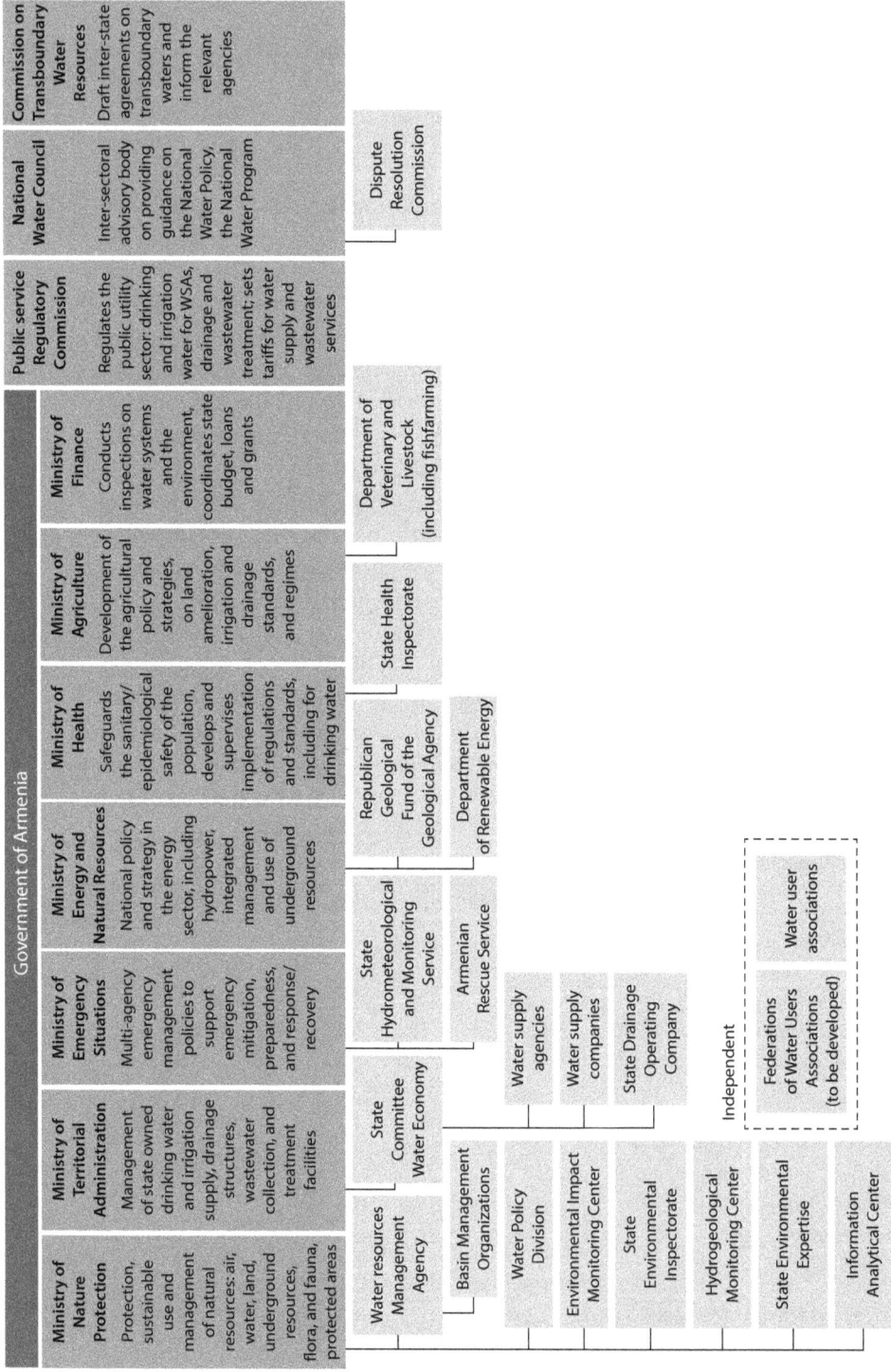

Government of Armenia

Ministry of Nature Protection
Protection, sustainable use and management of natural resources: air, water, land, underground resources, flora, and fauna, protected areas

Ministry of Territorial Administration
Management of state owned drinking water and irrigation supply, drainage structures, wastewater collection, and treatment facilities

Ministry of Emergency Situations
Multi-agency emergency management policies to support emergency mitigation, preparedness, and response/recovery

Ministry of Energy and Natural Resources
National policy and strategy in the energy sector, including hydropower, integrated management and use of underground resources

Ministry of Health
Safeguards the sanitary/epidemiological safety of the population, develops and supervises implementation of regulations and standards, including for drinking water

Ministry of Agriculture
Development of the agricultural policy and strategies, on land amelioration, irrigation and drainage standards, and regimes

Ministry of Finance
Conducts inspections on water systems and the environment, coordinates state budget, loans and grants

Public service Regulatory Commission
Regulates the public utility sector: drinking and irrigation water for WSAs, drainage and wastewater treatment; sets tariffs for water supply and wastewater services

National Water Council
Inter-sectoral advisory body on providing guidance on the National Water Policy, the National Water Program

Commission on Transboundary Water Resources
Draft inter-state agreements on transboundary waters and inform the relevant agencies

Dispute Resolution Commission

Water resources Management Agency

State Committee Water Economy

State Hydrometeorological and Monitoring Service

Armenian Rescue Service

Republican Geological Fund of the Geological Agency

Department of Renewable Energy

State Health Inspectorate

Department of Veterinary and Livestock (including fishfarming)

Basin Management Organizations

Water supply agencies

Water supply companies

State Drainage Operating Company

Water Policy Division

Environmental Impact Monitoring Center

State Environmental Inspectorate

Hydrogeological Monitoring Center

State Environmental Expertise

Information Analytical Center

Independent

Federations of Water Users Associations (to be developed)

Water user associations

Source: Based on UNDP/GEF 2013.

Notes

1. This is defined in the Code as "the quality and quantity of water that is required to satisfy present and future basic human needs, as well as to protect aquatic ecosystems and to secure sustainable development and use of that water resource."

2. Here "water system" is defined as those hydrotechnical structures related to the use of water resources causing the alteration of water flow or used to convey water resources, including (but not limited to) dams, dikes, embankments, canals, channels, wells, pipelines, pumping plants, purification plants, water outlets, spillways, aqueducts and water storage facilities, any machinery, appliances, or apparatus that are constructed, erected, or used for impounding, storage, conveyance, distribution, drainage, control or extraction of water, power generation, water treatment, water use, or rainfall collection.

3. "On the Procedures for Transferring State-Owned Irrigation Systems to Water User Associations by Free of Charge Use Rights and Ensuring Supervision (Control) of Those Systems by Owners" and "Transfer Agreement for Passing Over of State-Owned Irrigation Systems to Water User Associations by Free of Charge Use Rights."

4. As a result of recent reforms of the government of Armenia, two establishments (the State Hygiene and Antiepidemiological Inspectorate under the Ministry of Health and the State Labour Inspectorate of the Ministry of Labour and Social Affairs) were merged into a single agency—the State Health Inspectorate under the Ministry of Health.

References

Ministry of Nature Protection. 2013. *Feasibility of the Master Plan for Integrated Water Resources Management in the Six Water Basin Management Areas of Armenia.* Report funded by SOFINEX and prepared by SHER Ingénieurs-Conseils.

OECD (Organisation for Economic Co-operation and Development). 2013. *Facilitating the Reform of Economic Instruments for Water Management in Armenia.*

UNDP/GEF. 2013. "Updated Transboundary Diagnostic Analysis." Prepared for UNDP-GEF Project on Reducing Transboundary Degradation in the Kura Araks River Basin, Baku/Tbilisi/Yerevan, September.

USAID (United States Agency for International Development). 2008b. *Water Resources Atlas of Armenia.*

WRMA (Water Resources Management Agency). 2011. *Ten Years of Experience in Reforming Water Management Sector in Armenia: Towards the EU Water Framework Directive.*

CHAPTER 4

Emerging Challenges to IWRM

Strengthening Monitoring of Water Quantity and Quality

Key Messages

- Obtaining reliable, timely, good-quality, and publicly available data on water quantity and quality are precursors to a functioning integrated water management and planning system.
- Monitoring systems are vital to various planning and investment exercises, including in the issuance and compliance of water use permits (WUPs).
- Insufficient investment over decades in the monitoring infrastructure (including institutional capacity building) is evident, with opportunities to introduce new technologies and approaches to data collection, verification, and management.
- Improved coordination and harmonization across the various departments responsible for monitoring will be critical.

Introduction

Obtaining hydrometeorological information and data that are reliable, timely, of good quality, and publicly available is an essential precursor to good integrated water management and planning. Future investments cannot be fully prepared without a sufficient knowledge base on water resources in place. Moreover, day-to-day operations of the various water systems both for productive purposes (for example, irrigation, urban supply, environmental flows) and risk mitigation purposes (for example, flood warning) cannot be optimized without a robust near real-time monitoring network. Finally, management of the overall resource sustainability (for example, through permitting) and various competing pressures is only possible when data are being monitored over time and resource assessments updated regularly. In Armenia, there are several different agencies with responsibility for water monitoring (both quantity and quality, both surface water and groundwater). These include the State Environmental Inspectorate (SEI), the Hydrogeological Monitoring Center, the Environmental Impact Monitoring Center, and the ASHMS. Though duplication of efforts may not necessarily be inefficient, some rationalization and modernization is needed. Map 4.1 shows the locations of the water monitoring stations in Armenia.

Since Soviet days, very little investment has been devoted to strengthening the monitoring infrastructure. To enhance the current monitoring system, a comprehensive view must be taken. Over the last decade, investments in monitoring have been done in a piecemeal manner (a piece of equipment here, a piece of

Map 4.1 Water Monitoring Stations in Armenia

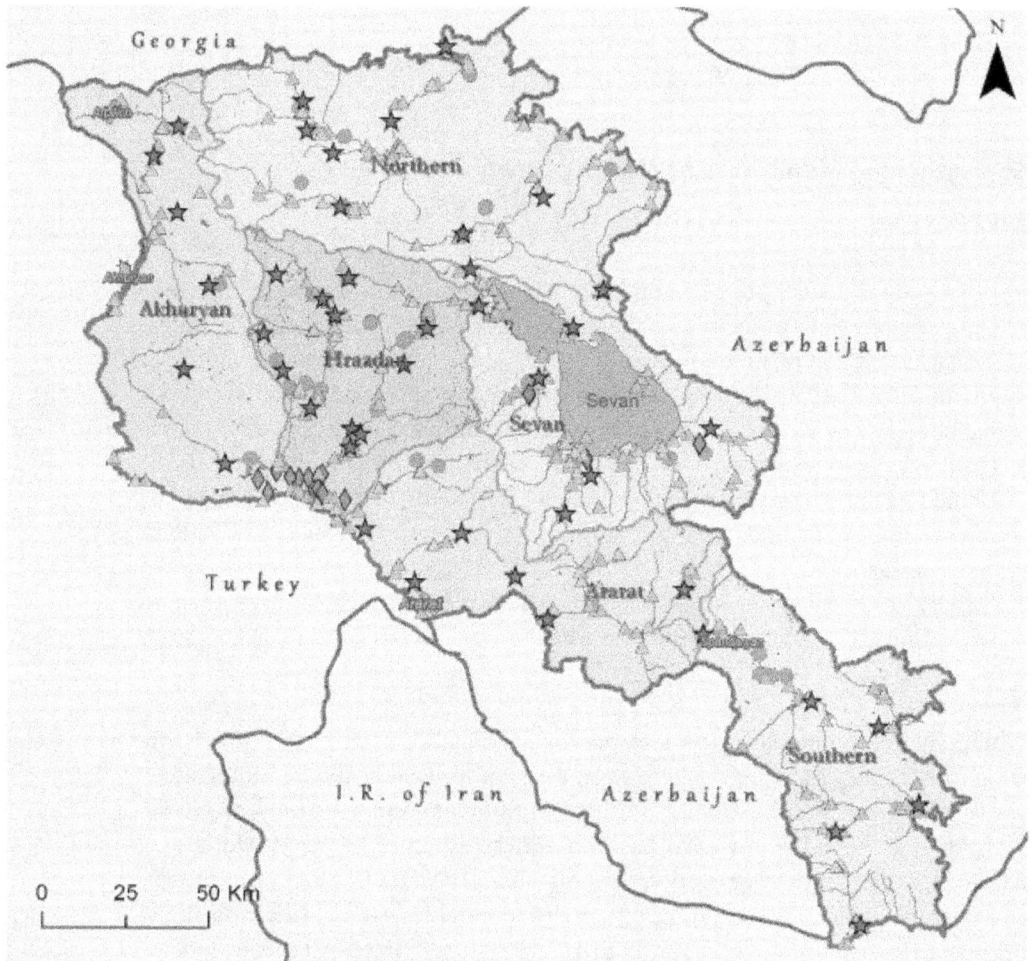

Rivers
Lakes and reservoirs
BMO

★ Meteorological station
▲ Water quantity station
▲ Water quantity station
◆ Groundwater well
⬤ Groundwater spring

Source: USAID 2008.
Note: A full-color version of this map may be viewed at http://www.issuu.com/world.bank.publications/docs/9781464803352.

Toward Integrated Water Resources Management in Armenia • http://dx.doi.org/10.1596/978-1-4648-0335-2

equipment there) with financing from outside donors. In most cases, the numbers of monitoring points could be expanded and the technologies used modernized (for example, through greater use of automated readers or real-time telemetry). Moreover, the quality of the monitoring infrastructure is poor and in many cases outdated (photo 4.1). Sharing of data among different agencies and access to

Photo 4.1 Pictures of Hydrological Monitoring Equipment

Source: © Vahagn Tonoyan. Used with permission, courtesy of Vahagn Tonoyan. Further permission required for reuse.

data by the public (through department websites) also remains very limited. It is generally accepted that such investments are highly economical. A study conducted by the World Bank (2006) examining the economic efficiency of investments in hydrometeorological services in Armenia revealed that an estimated US$19 in prevented losses (from hydrometeorological hazards) could be achieved for every US$1 spent.

A driving force in introducing standards and protocols for monitoring is with respect to the European Union Water Framework Directive (see appendix E). Under the Water Framework Directive, three areas of monitoring are required. First, surveillance monitoring is needed from a fixed system to identify long-term changes and trends and to inform future monitoring networks; second, operational monitoring is needed to help classify and observe identified water bodies that are at risk of failing to meet various objectives; and third, investigative monitoring is needed whereby particular problems and challenges are studied in depth.

Current Institutions Responsible for Water Monitoring and Status

Table 4.1 shows the institutions responsible for water monitoring in Armenia.

State Environmental Inspectorate (SEI). Though substantial progress has been made in terms of putting a water use permit (WUP) system in place, compliance and enforcement of these permits remain weak. The Ministry of Nature Protection designated the SEI responsible for the enforcement of WUP requirements. The SEI monitors the following: actual water extraction points or water supply systems (name and location); actual extracted water quantities (total, by quarters of the year and allowed by the permit); quantity of water actually used for various needs (drinking, municipal, irrigation, industry, rural supply, and other); quantity of actual water returned (total, returned to surface waters, quantity of polluted water, quantity of wastewater treated by mechanical, physical-chemical, and biological methods); description of the outflow, including volume of hazardous chemicals in the water used and returned (actual and maximum allowed discharge); content of harmful substances in wastewater discharged to water resources by basins, marzes (provinces), and communities (including total biological oxygen demand, ammonia nitrogen, nitrates, nitrites, phosphates, chlorides, sulfates, iron, copper, zinc, nickel, suspended substances); and wastewater discharge to water resources, categorized by basins, marzes, and communities. This information is collected and submitted by hard copy to the SEI regional offices. In order to determine the amounts of constituents that are

Table 4.1 Water Monitoring Institutions in Armenia

Monitoring function	Responsible agency	Ministry
Surface water quantity	ASHMS	Emergency situations
Surface water quality	Environmental Impact Monitoring Center	Nature protection
Groundwater quantity and quality	Hydrogeological Monitoring Center	Nature protection
Drinking water sources and quality	State Health Inspectorate	Health care
Water use and pollution discharge	State Environmental Inspectorate	Nature protection

being discharged to a water body and determine compliance, water samples are collected at source and analyzed. However, sampling and inspection is performed only once a year for prioritized sources and even less often for nonprioritized sources. Due to a number of factors (such as shortage of equipment), the inspection and sampling frequency is inadequate (UNECE 2010). Box 4.1 shows the structural and territorial units of the SEI.

Hydrogeological Monitoring Center. In general, the state of groundwater monitoring (both quantity and quality) is weak. Following Armenian independence in 1993, the Soviet Union Hydrogeological Expedition (of the then Geological Department) was closed and groundwater monitoring ceased. By 2006 the National Water Program reestablished a Groundwater Monitoring Program highlighting the priority given to the establishment and operation of a national reference monitoring network. With the support of the United States Agency for International Development (USAID) Water Program around this time, the existing groundwater monitoring points were reestablished and various assessments undertaken. Moreover, 73 monitoring points were identified for the reference network, including 49 natural springs, 22 borehole wells, and 2 groundwater wells. USAID support was provided to rehabilitate 69 of these 73 monitoring points. These were handed over to the Hydrogeological Monitoring Center in 2008. The observed parameters at the monitoring points include temperature, water level, and discharge.

Given the importance of groundwater (especially for drinking water purposes), a more robust monitoring network is required. Very few long-term time series data exist (some old records exist that have yet to be digitized). This would be critical in identifying trends across various aquifer subunits. Moreover, few pump tests (essential for estimating basic aquifer parameters such as hydraulic conductivity, storativity, or effective porosity) have been completed for the different stratigraphy layers. There is also currently no equipment for

---

Box 4.1 Units of SEI

A. Structural units of SEI

- Division of Water Resources Supervision
- Division of Atmospheric Air Supervision
- Division of Biodiversity, Soils, Wastes, and Hazardous Substances Supervision
- Division of Entrails and Surveyor Supervision
- Division of Forests Supervision
- Central Laboratory

B. Territorial units of SEI

- Yerevan Territorial Division
- Syunik Territorial Division
- Ararat Territorial Division
- Armavir Territorial Division
- Aragatsotn Territorial Division
- Gegharkunik Territorial Division
- Kotayk Territorial Davison
- Tavush Territorial Division
- Lori Territorial Division
- Shirak Territorial Division
- Vayots Dzor Territorial Division

---

automated monitoring of groundwater levels. Data collected are reported in simple text formats. There is no standardized data archiving, treatment and analysis software, or procedure established. With respect to groundwater quality, the Hydrogeological Monitoring Center collects the groundwater samples but outsources the laboratory analysis to the Environmental Impact Monitoring Center or to the Geological Laboratory under the Ministry of Energy, as the Hydrogeological Monitoring Center does not have its own laboratory. The center measures some major ions with some overlapping analysis done by the Environmental Impact Monitoring Center. These elements can only provide the basic characterization of the origin of groundwater and a gross indicator of pollution (not persistent pollutants). No multiparameter probes are available with the department.

Environmental Impact Monitoring Center.[1] This department is responsible for the collection of surface water quality data. The central office is in Yerevan, where the main laboratory is housed. Two regional offices exist at Vanadzor and Kapan. Staffing is about 52 people. After 1992 water quality monitoring was drastically reduced, making long-term time series data unavailable. In 1998, only 55 water samples were taken. State budgets have improved from 13 million Armenian drams (AMD) (US$32,000) in 2004 to 70 million AMD (US$170,000) in 2013. Moreover, with support from various donors (especially USAID and the European Union), modern laboratory equipment has been provided (including mass spectrometers and chromatographs). Since 2007, the Environmental Impact Monitoring Center has been in full operation, with 1,200 samples gathered from 131 observation posts (6–12 samples per year). At these posts, about 50 variables are measured. This includes pH, biological oxygen demand, chemical oxygen demand, conductivity, major ions, and some metals. Analysis is being conducted according to ISO standards or other international standards. The Environmental Impact Monitoring Center also has a specific procedure for data verification and validation. The center publishes monthly and annual printed bulletins in the Armenian language, which contain data on surface water quality.

Armenian State Hydrometeorological and Monitoring Service (ASHMS).[2] This department is the main authorized body for surface water quantity and meteorological monitoring in the country. The department is under the Ministry of Emergency Situations. The total staff is around 592, of which 382 have university education. The department currently operates and maintains 47 meteorological stations (including 6 high-altitude stations and 3 specialized stations), 2 agrometeorological stations, 7 hydrological stations, and 94 hydrological observation posts. Though, in general, the density of stations may be adequate, the majority of the hydrological and meteorological observation points are poorly equipped. More snowpack measurements are needed. All of the hydrological observations are using simple water level rulers affixed to a local structure (locations shown in photo 4.1). The data are collected at each point twice a day. At a small number of these points (7 out of 94) flow meters are used to record actual discharges. About 30 observations annually are made using flow meters. Few observation points have some kind of automation. More than 80 percent of the hydrological stations were installed before 1970. With the data collected from all observation posts, the ASHMS headquarters in Yerevan processes the obtained data and prepares annual reference books. All data are being stored in an electronic database, which is not available online.

In general, the implementation of these hydrometeorological monitoring programs has been hampered by low salaries, little capital investment, and low operational budgets. As of 2014, the budgets for the ASHMS were as presented in table 4.2. Within these budget constraints, measurements such as snowpack, sediments in reservoirs and lakes, cross-sectional surveys, and water turbidity have been curtailed. Also, taking into consideration the move toward a river basin management approach, the network may need to be reevaluated and observation points placed at critical points in the basins.

One critical function of the ASHMS is with respect to forecasting. This is critical for water management both in the short term (in the case of flooding or droughts) and in the longer term (for seasonal agriculture planning, for example). The current capacity of forecasts, which is very much dependent on the quality of the monitoring, is given in table 4.3.

State Water Cadastre Information System (SWCIS). The SWCIS was developed as a supporting tool for integrated water resources management (IWRM)

Table 4.2 Budget Indicators for ASHMS in Armenia in 2014

Budget categories	Expenditures as a percentage of budget
Total budget	**763 million AMD (US$1.84 million)**
For meteorological monitoring (47 meteorological stations)	450 million AMD (US$1.08 million); 59%
For hydrological monitoring (94 hydrological observation posts)	156 million AMD (US$0.38 million); 20%
For other activities (hydrogeophysical monitoring, scientific research, etc.)	157 million AMD (US$0.38 million); 21%

Source: ASHMS.
Note: AMD = Armenian drams.

Table 4.3 Forecasting Capacity of ASHMS

Time frame	Period/accuracy
Lead time of standard forecasts:	
Short range	4–5 days
Mid range	7–15 days
Long range	1 month and more
Lead time of household warnings (hours)	1–12 hours
Accuracy of forecasts and warnings:	
Short range	80–85%
Mid range	71–76%
Long range	65–70%

Source: ASHMS.

for the Water Resources Management Agency (WRMA). The Water Code defines this system as "a permanent operating system to keep comprehensive records of quantitative and qualitative indices on water resources, water intakes, watersheds, composition and quantities of materials and biological resources, as well as records of water users, WUPs, and water system use permits." Thus, this system aims to integrate all databases into a single framework to be accessed by a range of stakeholders (table 4.4).

The WRMA is in charge of consolidation and maintenance of all water resources and water system-related information in this official repository. The SWCIS consists of a centralized data warehouse, operated and maintained by the Water Resources Monitoring and Cadastre Division of the WRMA, that stores national-level water resources data (tabular and spatial) with customized applications capable of analyzing and processing the data, and database applications at stakeholder institutions with customized export tools for transferring data from each database to the data warehouse. Access to these water resources data via the Internet and for broader public consumption is not possible because the WRMA website has been down since 2008.

Conclusions

Though a network of surface water and groundwater monitoring exists (table 4.5), additional investment is needed to ensure adequate future IWRM planning. Some clear gaps are observed (on a case-by-case basis) with each agency. For instance, strengthened groundwater monitoring that involves improving the understanding of the various aquifer layers and changes over time is needed. Also, as per the European Union Water Framework Directive, more will need to be

Table 4.4 Stakeholder Institutions of the SWCIS and Available Data

Stakeholder institution	Available data
WRMA, Ministry of Nature Protection (authorized agency for State Water Cadastre)	Water use and wastewater discharge data
ASHMS, Ministry of Emergency Situations	Surface water quantity data
Environmental Impact Monitoring Center, Ministry of Nature Protection	Surface water quality data
SEI, Ministry of Nature Protection	Actual water use and wastewater discharge data
Republican Geological Fund of the Geological Agency, Ministry of Energy and Natural Resources	Inventory of groundwater resources
Hydrogeological Monitoring Center, Ministry of Nature Protection	Groundwater quality and quantity data
State Committee on Water Systems under the Ministry of Territorial Administration	Water systems used for drinking water supply, irrigation water intake, drainage structure-operating organizations, and water user associations
State Inspectorate, Ministry of Health	Drinking water quality monitoring, water monitoring of open reservoirs, violations of sanitary norms

Source: European Union 2011.

Table 4.5 Summary Information on Surface Water and Groundwater Monitoring Points

Basin	Area (km²)	Surface water quantity gauging stations		Surface water quality sampling points		Groundwater springs and wells
		no.	km² for 1 station	no.	km² for 1 station	
Akhuryan	5,044	17	297	14	360	14
Ararat	4,460	13	319	16	279	8
Northern	7,068	23	307	25	283	39
Sevan	4,806	14	339	22	216	3
Hrazdan	3,881	16	243	33	118	1
Southern	4,484	9	498	21	213	8
Total	**29,743**	**92**	**334**	**131**	**245**	**73**

Source: European Union 2011.

done with respect to water quality, including enhancing hydrobiological monitoring and monitoring of the directive's pollutant priorities. In many cases, equipment could also be modernized with greater automation and real-time monitoring added. This may include the use of integrated monitoring approaches such as joint water quantity and quality stations. A more comprehensive strategic analysis of the monitoring requirements of the country and the capacity requirements to maintain such systems is needed. Though the SWCIS is meant to comprehensively consolidate this information and make it available online for a broad audience, this has yet to be achieved. Further strengthening of data-sharing mechanisms, particularly between the ASHMS and the WRMA, would be helpful. Overall, improved coordination and harmonization of surface water and groundwater quantity and quality monitoring activities will be critical.

Weakness in River Basin Management Planning

Key Messages

Despite the various initiatives supported by the donor community, the water sector in Armenia still faces many challenges in terms of river basin management planning.

- Needed skills and data to carry out modeling and planning work are not adequately available within the basin management organizations (BMOs).
- The current river basin planning template relies heavily on the European Union Water Framework Directive and focuses primarily on achieving good ecological status of water bodies.
- Broader intersectoral planning that takes into account municipal, agriculture, energy, and environment linkages and the various departments responsible is not sufficient.
- Completed river basin management plans (RBMPs) have yet to be adopted by the government. Government endorsement of such plans is needed to ensure

that all levels of government have a consistent planning vision and a clear prioritization of future investments.

- Analysis and knowledge on what would be the best allocation (both in economic and efficiency terms) for the different water users in the basin is needed.
- Lack of State-level budget is likely to undermine ongoing planning efforts and
the full participation of BMOs in river basin planning.

Introduction

According to the Water Code, basin management authorities are called upon to
develop river basin management plans (RBMPs). The RBMP is a comprehensive
document that describes the management and conservation activities to be
implemented within a river basin in order to achieve the objectives laid out in
the Water Code. In line with international best practice, the Water Code strongly
supports water resources (both surface water and groundwater) planning to be
done at the level of the basin. Moreover, the Water Code highlights that RBMPs
need to give sufficient attention to intersectoral balance among community, irrigation, energy, industry, and ecological uses. RBMPs are to be developed with the
full participation of stakeholders.

The 2006 National Water Program included specific provisions for the development of these plans. Development and implementation of RBMPs will be an
essential guiding framework for the basin management organizations (BMOs).
Already several draft RBMPs have been developed or are in the process of development (Debed, Aghstev, Marmarik, Vorotan, Meghriget, Arpa, Akhuryan,
Metsamor river basins). The government has yet to officially adopt, fund, or
implement any of these plans.

Recent Water Resources Planning Efforts

Various river basin planning efforts have been completed or have been ongoing
since 2007 (USAID 2012b). Following the requirements established in the
Water Code, a model guideline for the formulation of the RBMP was developed
in 2008 with the support of USAID. The model guideline was based on the
principles of IWRM and the provisions of the European Union Water Framework
Directive (box 4.2).

As shown in table 4.6, all of these RBMP efforts have been undertaken with
the financial support of the donor community. In terms of coverage, almost all
river basin management areas of Armenia have been covered. Map 4.2 indicates
the geographic coverage of the completed or ongoing RBMPs and respective
donors. None of the plans has been approved by government, and therefore they
have no binding legal basis. Government endorsement of such plans is needed to
ensure that all levels of government have a consistent planning vision and a clear
prioritization of future investments.

The results of these early planning efforts (2008–10) have certainly provided valuable lessons and information for water resources planning in
Armenia. Moreover, given the need to adopt further the existing model guideline to local conditions, in 2011 a protocol describing the elements that should

Box 4.2 RBMPs under the European Union Water Framework Directive

The European Union Water Framework Directive establishes that each member country has to produce and publish river basin management plans (RBMPs) by 2009 for each river basin district, including the designation of heavily modified water bodies, while encouraging the active involvement of all interested parties in their development and implementation. According to the directive, the RBMP is primarily intended to record the current status of water bodies within the river basin district; set out, in broad terms, what measures are planned to meet environmental objectives; and represent the main reporting mechanism to the European Commission and to the public. The plan should also summarize how the objectives set for the river basin (ecological status, quantitative status, chemical status, and protected area objectives) will be reached within the time scale required.

Table 4.6 River Basin Planning Efforts between 2008 and Present

Date completed	Financial support	River basin management area and status
2008	USAID	Application of the model guideline to the Meghriget River and its tributaries following into the Araks River.
2010	UNECE	Baseline conditions for and pressures facing IWRM in the Marmarik River basin, setting desired conditions for water uses and functions, and identification of measures to achieve desired conditions.
2010	European Union	Formulation of draft RBMPs for Aghstev and Debed Rivers based on Water Framework Directive requirements, and identification of water bodies at risk in terms of their ecological status and potential restoration measures.
2013	UNDP/GEF	Arpa RBMP being developed within the framework of the UNDP/GEF Reducing Transboundary Degradation in the Kura-Araks River Basin Project.
In progress	USAID	RBMPs for the Southern basin management area, comprising the Vorotan, Meghriget, and Voghji Rivers and their watersheds, are in progress. The draft final report on the Vorotan is ready, and includes comments from the stakeholder institutions and ministries. It will be discussed with the public in April 2014.
In progress	European Union	The Akhuryan-Metsamor RBMP is being prepared within the European Union Environmental Protection of International River Basins Project.

Note: GEF = Global Environment Facility; UNECE = United Nations Economic Commission for Europe; UNDP = United Nations Development Programme.

be included in each RBMP was adopted by the government. The Content of Model Water Basin Management Plan Protocol, which draws heavily on the European Union Water Framework Directive, currently provides the basis for the development of RBMPs in the country. The protocol, however, is not fully consistent with the Water Code, which specifically stipulates that the basin plans "shall balance the interconnected relationship of all water users, including communities, power generation, industry, agriculture, and environment." The European Union Water Framework Directive takes a narrower approach and focuses mainly on the protection of the aquatic ecosystem (USAID 2012b). Under the ongoing Clean Energy and Water Program, USAID has been

Map 4.2 Coverage of RBMPs in Armenia

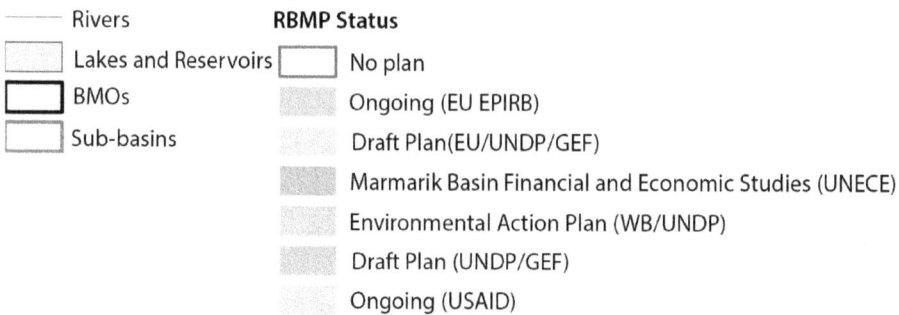

	Rivers	**RBMP Status**
	Lakes and Reservoirs	No plan
	BMOs	Ongoing (EU EPIRB)
	Sub-basins	Draft Plan(EU/UNDP/GEF)
		Marmarik Basin Financial and Economic Studies (UNECE)
		Environmental Action Plan (WB/UNDP)
		Draft Plan (UNDP/GEF)
		Ongoing (USAID)

Note: A full-color version of this map may be viewed at http://www.issuu.com/world.bank.publications/docs/9781464803352.

supporting the government of Armenia in the development of a further improved framework. Table 4.7 provides a comparative analysis of the content of the current and proposed model RBMP.

While the USAID-proposed revised model RBMP is a step in the right direction toward improving the structure of the RBMP, there remain additional adjustments to consider in the overall planning framework. This includes ensuring that the framework addresses issues of competition between different users to ensure efficient water use, the balancing of available water supply and demand during the dry season, effective flood management during the wet season, climate change resilience and adaptation, river basin transfers, and protection and conservation. Most importantly, the RBMPs must be clear on how various planning and investment choices in the water sector (across all subsectors) link with other economic sectors in the Armenian economy, such as energy, agriculture, and mining. These planning efforts are also opportunities for the government of

Table 4.7 Comparative Analysis of Content of Current and Proposed Model RBMP

Component	Content of current model RBMP	Content of proposed model RBMP
Issue Identification	Main description of the river basin	River basin characterization
	Identification of current conditions and functions in the river basin	Identification of current status of water use and activities in the river basin
	Assessment of natural and anthropogenic impacts on water, including assessment of climate change impacts	Assessment of natural and anthropogenic pressures and impacts on water resources of the river basin
Priority Setting	Identification of desired conditions and functions in the river basin	Identification of desirable status of water use and activities in the river basin
	Classification of water bodies of the river basin delineated according to management peculiarities	Classification of water bodies delineated according to peculiarities of water resources management
	Definition of the ecological flow of water bodies in the river basin	Calculation of environmental flow for water bodies in the river basin
	Identification of measures toward achieving the desired conditions in the river basin	Program of measures for achieving desirable status
	Identification of measures to mitigate the possible consequences of and prevent emergency situations in the river basin	Defining measures for prevention, mitigation, and elimination of consequences of emergency situations
	Assessment of water use demand in the river basin according to sectors	Assessment of water use demand by sectors in the river basin
	Water resources improvement scenarios according to sectors	Water resources improvement scenarios by sectors in the river basin
Economic and Financial Considerations	Preliminary financial assessment of identified measures	Preliminary economic and financial assessment of measures identified
	Assessment of existing financial deficit in the river basin according to sectors	Assessment of financial deficit by sectors in the river basin
Stakeholder Involvement	Involvement of public and stakeholder institutions in decision making	Involvement of public and interested agencies in decision-making process
Implementation and Monitoring		Provisions for continuous implementation of the RBMP

Source: USAID 2012b.

Armenia to examine water resource allocation scenarios across the full range of departments involved.

Moving forward, the government will need to invest budgetary resources in these multi-departmental basin planning efforts. To date, all basin planning efforts have been supported by external donors. This potentially poses two problems. First, the scope of the basin plans will be dictated, to some degree, by the donors and their interests. Second, the basin plans have relied heavily on the use of external consultants (often internationals) with limited long-term involvement with the BMOs. The BMOs are the owners of these RBMPs and ought to be driving the diagnostics of the basin and the participatory planning process. A case in point is the approach adopted for the formulation of the Akhuryan-Metsamor RBMP.[3] According to the recent call for proposals under the Environmental Protection of International River Basins Project, the basin plan will be consistent with the requirements of the European Union Water Framework Directive, as the program is funded by the European Union. As the objective of the project is to improve water quality, it is very likely that the primary focus of the basin plan will be on achieving a desired water quality status and not on broader water resources planning issues.

Conclusions

Despite these various basin planning efforts, so far no basin plan has been completed and approved by the government. Government endorsement of such plans is needed to ensure that all levels of government have a common planning vision and a clear prioritization of future investments in the sector. As a result, analysis of and knowledge on what would be the best allocation (in economic and efficiency terms) for the different water users in each basin are not available. This is despite the fact that water permit and allocation decisions are routinely being made. Currently, the planning of irrigation, water supply, and hydropower investment programs, which are managed at the central level, has limited relationship with the RBMPs. Thus, a clear disconnect exists between the basin plans and the sector programs and budgets. In moving forward, the government will need to invest in multidepartmental basin planning efforts. The RBMPs must be clear on how various planning and investment choices (across all subsectors) contribute to the overall economy of the country.

Strengthening the Water Permit System

Key Messages

- The permitting process is the main regulatory tool for IWRM.
- The Water Resources Management Agency (WRMA) is the agency responsible for issuing permits. This function is expected to be devolved to the basin management organizations (BMOs) as their capacities develop.
- Ensuring compliance of water permits is currently insufficient due primarily to lack of resources and agency capacity.

- Compliance involves a monitoring function and an enforcement action function. These roles and responsibilities have been separated under the current legislative framework.
- Greater cooperation (perhaps legislated) on inspection and enforcement is needed between the WRMA and the SEI to reduce duplication and overlap in functions and increase monitoring efficiency.
- Compliance history can be made a more explicit part of the permitting process.
- Compliance promotion (and more reliance on self-monitoring) is weak.
- Categorizing the size of water uses and pollution discharges, including establishing a limit for which a water use permit (WUP) is not required, would help to enhance agency efficiency.
- Greater public participation in the permitting process may be envisioned.

Background

Following the 2002 Water Code (chapter 4, articles 21–37), the Law on Environmental Oversight (2005), and the Law on Preparing and Implementing Inspections in Armenia (2000), the legal provisions for water use permitting have been established. Though the legislation provides the broad contours of how WUPs are to be applied, including the application process, contents, and criteria for review, it does not provide adequate guidance on compliance assurance and enforcement. Moreover, the full effectiveness of the water permit function is not possible due to lack of human, technical, and financial resources for compliance and weak public participation input (USAID 2007). In May 2011 the government of Armenia adopted Decision 677-N on making changes to Decision 218-N, dated March 7, 2003, on establishing standard forms of WUP and approving WUP forms. These changes improved the existing procedures on issuance of WUPs by stipulating additional procedures for issuance and extension of duration of the permit, and for providing hydrogeological data obtained from the Hydrogeological Monitoring Center of the Ministry of Nature Protection. These changes also allow for electronic submission of applications for permits and issuance of permits. Map 4.3 shows the locations of WUPs as of 2008.

The WRMA has the primary responsibility of issuing WUPs.[4] More specific guidelines on the permitting process were prepared in 2003 to support the WRMA. The guidelines—which have not been legislated—provide detailed descriptions of the permitting process, the rights of the applicant, evaluation criteria, public notification and input measures, and other useful guidance (both for applicants and for the WRMA). The steps given in the guidelines are shown in figure 4.1. The water use application contains a basic description of the proposal for water use and an analysis of its possible effects on water resources, ecosystems, protected areas, and people. The WUP applies to withdrawals from surface water and groundwater and controls the amount of extraction and the discharge quality. The WRMA determine the necessity of performing an environmental impact assessment. Public input is solicited at various points in the permitting process. No other water management agencies (for example, the State

Map 4.3 Locations of Water Use Permits

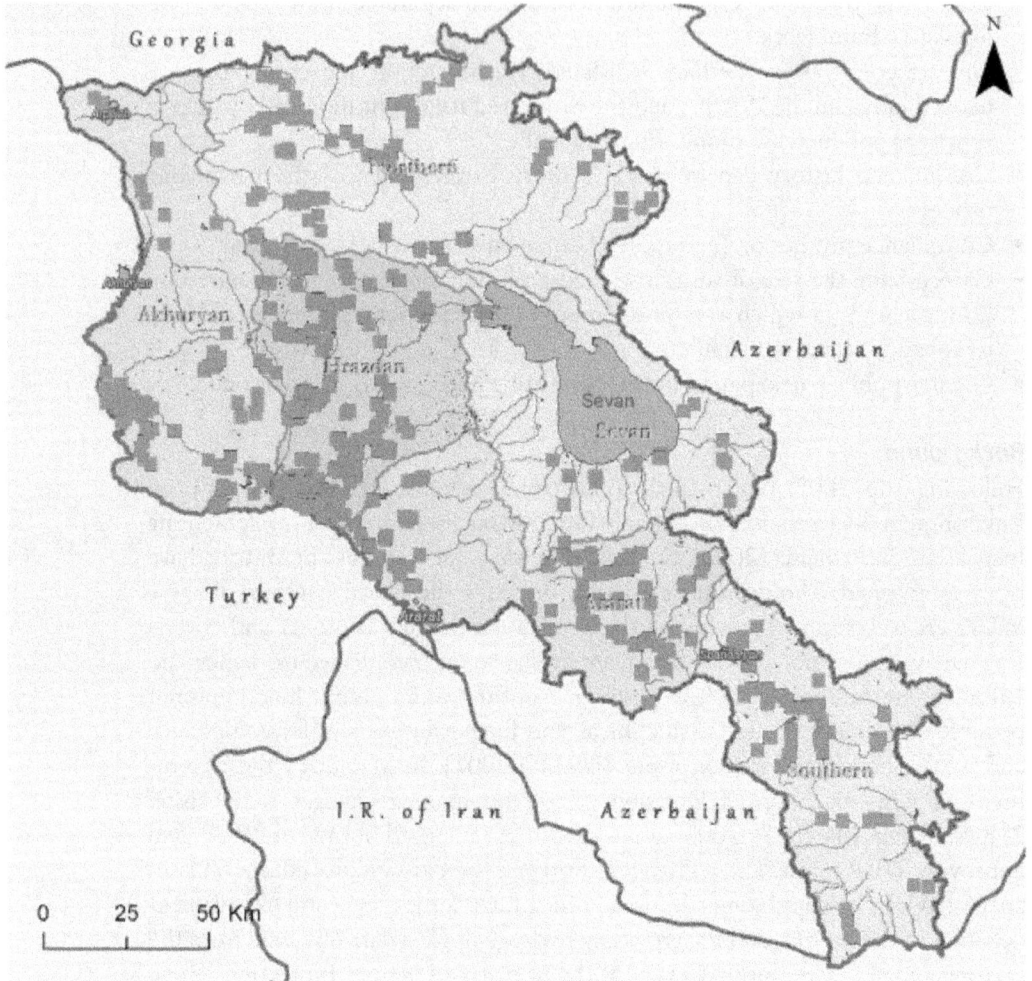

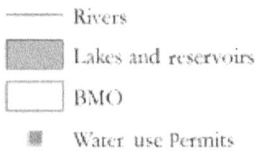

Source: USAID 2008.
Note: A full-color version of this map may be viewed at http://www.issuu.com/world.bank.publications/docs/9781464803352.

Committee on Water Economy) have a formal role in the permitting process. BMOs in the future are expected to take over this role.

Typical contents of a water permit application are the following:

- Address of the water user;
- Water use site;

Figure 4.1 Application and Issuance Processes for Water Use Permit

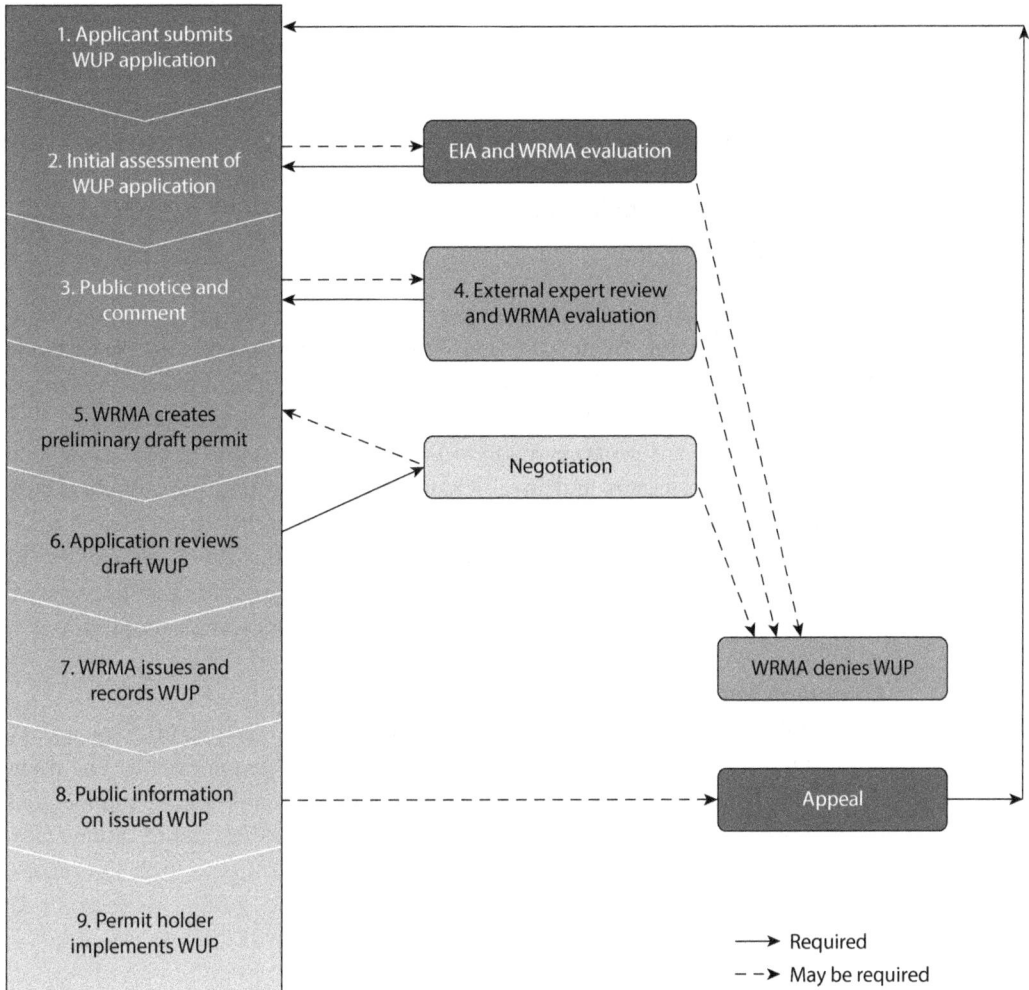

Source: USAID 2005.
Note: WUP = Water Use Permit; WRMA = Water Resources Management Agency; EIA = Environmental Impact Assessment

- Water abstraction site;
- Purpose of water use;
- Volume of water use, including from surface water resources and groundwater resources;
- Water use period and regime;
- Control mechanism to ensure the conditions of the WUPs;
- Allowable volumes of wastewater discharged into water resources or their watersheds;
- Description of discharge;
- Data on marginal allowed discharges of hazardous substances;
- Water standards and related information;

- Special measures that will be applied to promote efficient water use and improve water quality, wetlands and other important habitats, and related biodiversity;
- Corresponding requirements for water use calculation, monitoring, registration, and correction;
- Corresponding guarantees in case of causing damage to water resources;
- Payments associated to water use and payment schedule;
- Number of days within which the water use is subject to registration at the State Water Cadastre.

The Water Code (article 35) provides that a WUP holder may sell or otherwise transfer, in whole or in part, their WUP to a third party. The Water Code also provides for the inheritability of WUPs. There is debate as to whether this is appropriate (USAID 2007).

Despite significant progress made in WUP procedures during the last several years, there are still gaps and insufficiencies in the existing procedures. These include primarily (a) lack of regulations defining the marginal quantities of water use that do not require a WUP (as required by article 22 of the Water Code); (b) weaknesses in compliance and enforcement; and (c) low rate and fragmented nature of implementation, despite the formal public notification requirement.

Compliance and Enforcement

The Ministry of Nature Protection designates the SEI as the primary agency responsible for the enforcement and oversight of WUP requirements. The mandate of the SEI goes beyond enforcement of the permit requirements and includes enforcement of other environmental legislation related to air quality, biodiversity, soils, waste, and hazardous substances (for example, permits to emit into atmosphere, waste disposal permits, and permits to use and protect underground). Supported by the Law on Environmental Oversight, the SEI is authorized to:

- Enforce restrictions on the illegal and improper use of water resources;
- Enforce compliance with the requirements of water use, whether or not permitted;
- Establish and monitor parameters on the pollution of water resources in excess of the established limits;
- Enforce restrictions on water use in catchment basins;
- Enforce the rules related to maintaining the State registration of water resources;
- Ensure compliance with the requirements for the placement of landfills, dumps, cemeteries, and other facilities that may have an indirect harmful impact on water resources;
- Ensure compliance with the requirements for work within a specified distance of water resources that may affect or significantly impact the water resources;
- Ensure compliance with the requirements for recreational use of water resources;

- Ensure compliance with the national water program defined norms, limits, and restrictions for the use and preservation of water resources.

Moreover, the Law on Environmental Oversight provides guidance on the procedures for sanctions for noncompliance (including the fine and penalty structure), the use of inspections and examinations, and the rights and responsibilities of the water user permit holders. Currently, the SEI is not involved directly in the actual permitting process (for example, by providing compliance history information as part of the review process).

The SEI has 220 professional employees and 30 support staff. The challenges that the SEI faces in effectively performing its duties are well documented (USAID 2007). Due to insufficient funding, inadequate laboratory infrastructure and equipment, lack of sufficient technical tools and equipment (for example, computers and local and wide area networks), lack of appropriate trained personnel (such as environmental engineers), poor data transfer protocols, and difficulties with the recruitment procedures, the activities of the SEI related to water are largely impaired, and for most of the water users, there is no reliable information on whether they comply with the provisions mentioned in the WUPs or not. Thus, currently the level of compliance assurance and enforcement of the WUP conditions is insufficient.

It should be noted that this environmental oversight function is in contrast to the function that the WRMA and BMOs play. If the WRMA or BMO identifies a noncompliant permit holder, they write a detailed report and send it to the SEI. However, cooperation and information exchange among the BMOs, the SEI, and the Environmental Impact Monitoring Center is minimal. This is essential to the overall permit system, given that the SEI is required to measure the direct discharges, the Environmental Impact Monitoring Center is required to measure the quality of the receiving surface water body, and BMOs are in charge of State supervision of uses. Thus, to enhance this oversight, more coordinated action between agencies is needed. Legislative clarification of each agency's responsibility in relation to compliance assurance may be needed.

With regard to the Water Code, each WUP shall also clearly identify an adequate means of recording, monitoring, reporting, and verifying the water use and discharge by the permit holder. Thus, the water user is obligated to provide some level of self-monitoring. Currently, Parliament is reviewing the Law on Implementing Self-Monitoring of Requirements of the Environmental Legislation. Article 13 of the draft law outlines the self-monitoring and reporting procedures for discharging substances that pollute water resources and for using water in the production cycle. The self-monitoring is applicable to water users annually discharging over 1 tonne of biological oxygen demand or suspended particles or 100 kilograms of heavy metals, as well as to water users that abstract over 10 liters of water per second (except for hydropower and fishery sectors). Thus, strengthening this system of self-monitoring will help reduce the burden of compliance assurance on the public agencies and the response time to any environmental problems that may arise.

Public Participation in Issuing Water Use Permits

To provide greater transparency and public participation in the decision-making process, public notice and environmental impact assessment requirements are part of the WUP application process. These are highlighted in the Water Code. Article 5 (on basic principles of management, use, and protection of water resources and water systems) recognizes the importance of public participation and awareness in the process of management and protection of water resources. Article 20 (on public participation) lists the items that are subject to public notice (for example, draft RBMPs, pending WUPs, draft water tariff strategy, and draft water standards). Article 106 (on participation of nongovernmental organizations and citizens in the protection of water resources and water systems) defines the role of nongovernmental organizations and citizens in this process. The Water Code also provides a mechanism by which the public may file a complaint on a WUP decision. Finally, the specifics on public consultation are fixed in government Resolution 217-N of March 7, 2003, on Approving the Procedures for Ensuring Public Notification and Transparency of Documents Developed by the Water Resources Management and Protection Body, and its subsequent amendment of March 3, 2005.

The current permit guidelines require public notification and comment at the initial review process and then after a final decision. They do not require public notification on the proposed decision. Hence, potentially affected stakeholders do not have an opportunity to study any reports on the impact of the proposed application and the proposed permit conditions before the final decision. Thus, the procedures for public notification and appeal need improvement. The WRMA should provide ample time to the public to study the results of any impact studies, the justification of the proposed decision, and the proposed conditions for the permit. A few weeks after the provision of this kind of information, a public hearing could be organized by the BMO.

Conclusions

WUPs are one of the key tools for management and allocation of water resources in the country. Improved implementation of the WUP system is still constrained by (a) deficiencies in permitting regulations (for example, free water use, which is described in article 22 of the Water Code, is not defined yet, and the potentially affected stakeholders and public do not have an opportunity to study any reports on the impacts of the proposed application); (b) insufficient cooperation among agencies of the Ministry of Nature Protection in the processes of issuance of permits and the assurance of compliance with permit conditions (particularly between the WRMA/BMOs and the SEI); and (c) capabilities and resources of agencies and their staff (for example, the list of pollutants is too long, and most of the pollutants in the list cannot be effectively monitored by the SEI). To improve the permit system, categories may be defined for small, medium, and large water use and pollution discharges. Categorization is needed as large withdrawals require a more comprehensive and complex impact study and public notification process than small withdrawals. Also, establishing a limit below which a WUP is not required would help to reduce the agency burden. This would give the WRMA/BMO more time to process and focus on those permit applications that have a significant impact (and strategic impact) on the local water system and its users.

The Future of Ararat Valley

Key Messages

- The agriculture and fishery sectors are of strategic importance to the Armenian economy; Ararat valley is the largest agriculture and fish farming zone.
- Since 2006, a large number of fish farms have been established in the Ararat valley, in part due to the rich supply of artesian groundwater of high quality and low cost.
- Due to both overissuance of water user permits and overabstraction of groundwater resources above permitted levels or without water user permits, artesian groundwater resources are sharply depleting.
- Due to artesian groundwater depletion, the conflicts with other artesian groundwater uses—irrigation, domestic, industrial, and cooling waters—are increasing.
- Increasing water discharge from fish farms is overloading the drainage network, causing higher operation and maintenance costs, waterlogging, soil salinization, and alkalization.
- Several measures are being put in place. However, coordinated action across several ministries is required.

Background

The Ararat valley is the largest plain in Armenia (photo 4.2). It is divided into two parts—the northern part in Armenia and the southern part in Turkey. In Armenia,

Photo 4.2 Ararat Valley

Source: ©World Bank/Ju Young Lee. Used with Permission; further permission required for reuse.

the Ararat valley covers two administrative territories (parts of Ararat and Armavir marzes) and three BMOs (Akhuryan, Hrazdan, and Ararat). The valley is located at 800–1,000 meters above sea level and occupies an area of about 1,300 square kilometers on the Armenian side (map 4.4) (USAID 2014). The soil is fertile and the climate conducive to crop production. The Ararat valley is the largest agricultural zone in Armenia, providing up to 40 percent of the agricultural GDP (USAID 2012a). Various crops for export and local consumption are produced, including wheat, vegetables, grapes, and other fruits.

Map 4.4 Location of Ararat Valley

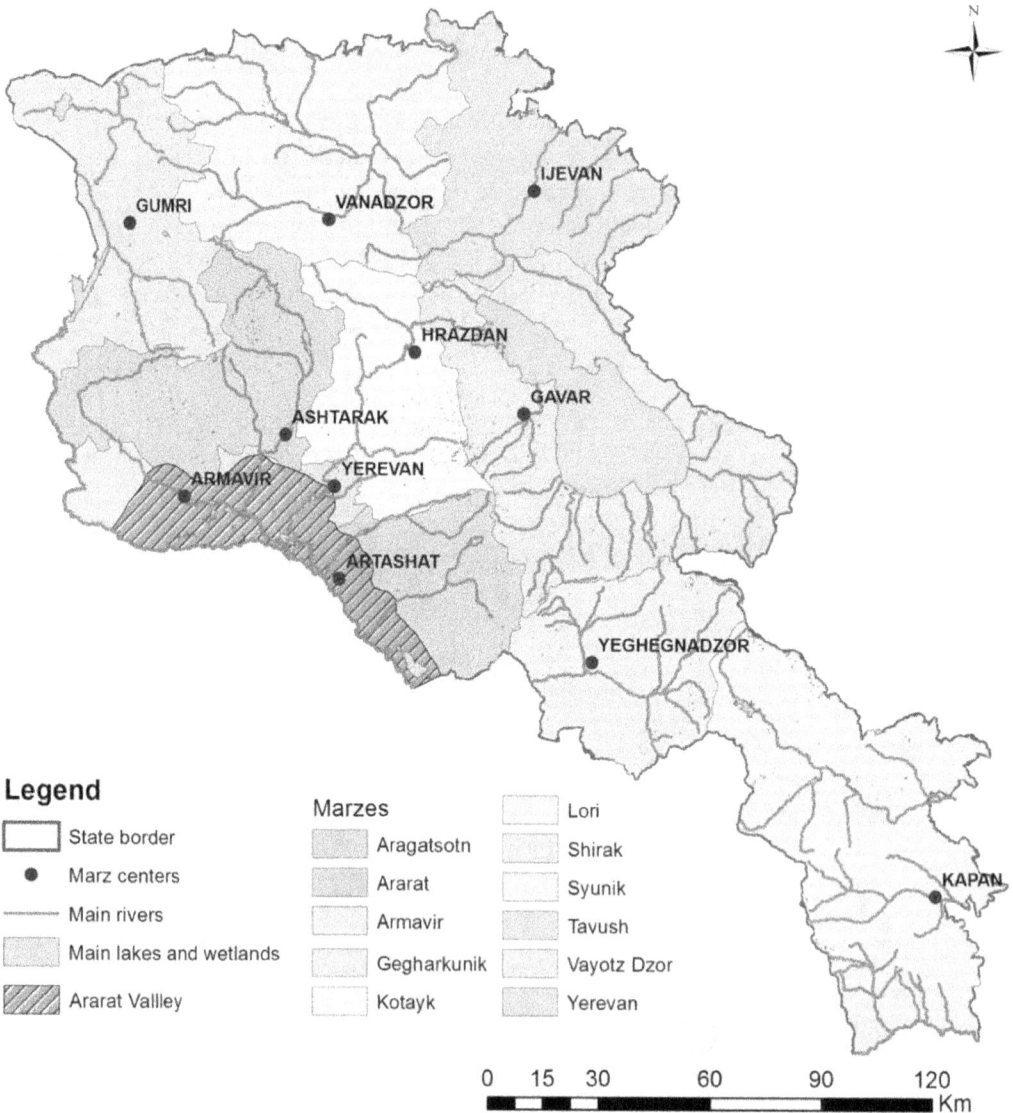

Legend

- ☐ State border
- ● Marz centers
- ⎯ Main rivers
- ☐ Main lakes and wetlands
- ▨ Ararat Vallley

Marzes

- Aragatsotn
- Ararat
- Armavir
- Gegharkunik
- Kotayk
- Lori
- Shirak
- Syunik
- Tavush
- Vayotz Dzor
- Yerevan

0   15   30   60   90   120
Km

Source: USAID 2014.
Note: A full-color version of this map may be viewed at http://www.issuu.com/world.bank.publications/docs/9781464803352.

The Ararat valley is rich with high-quality artesian groundwater, which is suitable for drinking purposes without additional treatment and comprises a strategic reserve of drinking water for the country. The artesian groundwater is at a depth of about 100–180 meters and is under high pressure.[5] This resource has historically been used for drinking and irrigation purposes. In recent years, fisheries have become one of the major water users (USAID 2012a).

During the last decade, development of private fish farms in the Ararat valley has significantly intensified due to the availability of low-cost, high-quality artesian groundwater, which supports year-round industrial production of fish. The number of fish farms in the valley has increased from just a few in the 1980s to 190 in operation in 2013 (109 in Ararat marz and 81 in Armavir marz) (USAID 2014). In 2013, the aquaculture fish production was 11,520 tonnes in Armenia. According to the Ministry of Agriculture, the potential for fish production is as much as 25,000 tonnes per year (FAO 2011). Given the continued domestic and international demand for fish,[6] the fish industry is of strategic importance to the country. However, along with the uncontrolled expansion of fish farms, there are growing concerns over the unsustainable use of groundwater. Box 4.3 presents further information on the economics of fish farming in Armenia, while figure B4.3.1 illustrates the growth in fish farming; box 4.4 gives data on a sample fish farm, and photo B4.4.1 shows pictures of the farm.

Groundwater Uses and Depletion in Ararat Valley

Groundwater abstraction currently exceeds the sustainable yield. The renewable level of groundwater use in the Ararat valley has been assessed by various authors. In 1984, the State Committee on Reserves approved a safe annual yield of artesian groundwater resources of 1,785 million cubic meters (MCM) per year (1,094 MCM from wells and 691 MCM from natural springs) (USAID 2014). According to various experts, these levels are still reasonable and do not need to be reassessed (USAID 2014). Even before the intensive development of the fish farming industry (around 2007), groundwater use in the Ararat valley already exceeded this renewable level (figure 4.2). According to the Hydro Institute inventory of wells and springs in the Ararat valley in 2007, there were 1,986 wells abstracting 1,151 MCM per year (USAID 2013). For fish farming, 299 wells were abstracting 401 MCM per year.

Fish production was included in the list of priority development programs in 2008 and thus more WUPs were issued (USAID 2014). In the period 2008–13, WUPs were issued for 274 new wells with a total discharge of 735 MCM per year, of which 202 were for fish farming (table 4.8). As a result, the actual groundwater abstraction from fish farms has increased by 719 MCM per year. It should be noted that this increase is concentrated mainly in the Ararat valley (Masis in Ararat marz and Echmiadzin in Armavir marz).

Considering both groundwater uses with and without WUPs, the total groundwater use in 2013 was 1.6 times the level approved by the State Committee on Reserves. Groundwater use by fish farms alone exceeded this level (figure 4.5). Though the actual abstraction from permitted wells for all

Box 4.3 Economic Value of Fish Farming in Armenia

In 2013, there were 335 fish farms officially registered in Armenia, of which 250 are operating. The total water area of fisheries in Armenia is 3,542 hectares. Fish farms, in terms of number and water area, are mostly concentrated in the Ararat Valley. As shown in figure B4.3.1, aquaculture fish production has been growing since 2007. It was 8,850 tonnes in 2012, 11,520 tonnes in 2013, and is estimated to reach 13,800 tonnes in 2014. Fish exports are also increasing: 1,800 tonnes in 2012 and 2,400 tonnes in 2013. In terms of value, by 2012 fish production was valued at around US$36 million (15 billion Armenian drams [AMD]). This means on average over US$4,000 (1.7 million AMD) per tonne of fish and over US$10,000 (4.2 million AMD) per hectare of fish farm.

Figure B4.3.1 Aquaculture Fish Production in Armenia

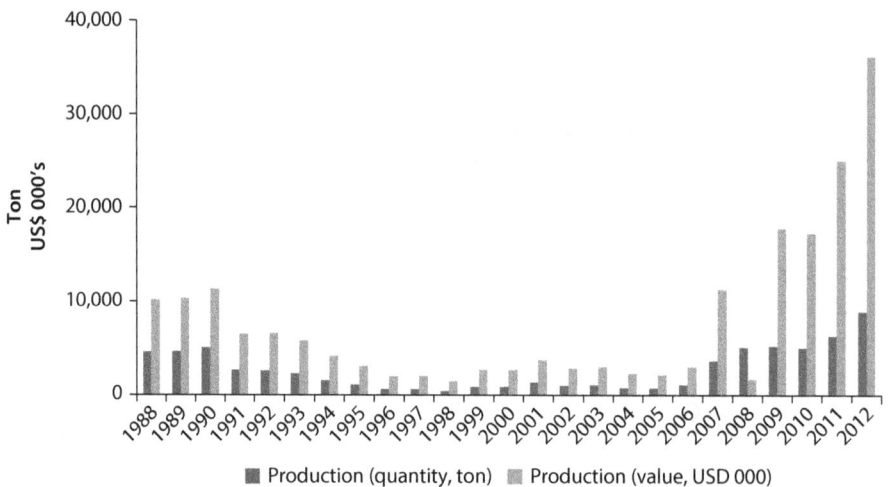

Source: FAO FishStat database: http://www.fao.org/fishery/statistics/en.

Source: FAO FishStat database: http://www.fao.org/fishery/statistics/en. FAO 2011. Ministry of Agriculture.

Box 4.4 Sample Fish Farm in the Ararat Valley

Basic facts

• Location: Dashtavan, Ararat marz
• Operation: since 2007
• Area: 3 ha (slightly larger than the average small fish farm size)
• Fish species: sturgeon and trout, which is indigenous from Lake Sevan
• Production: 50–60 tonnes/yr

box continues next page

Box 4.4 Sample Fish Farm in the Ararat Valley (continued)

Photo B4.4.1 Pictures of Sample Fish Farm

Source: ©World Bank/Ju Young Lee. Used with Permission; further permission required for reuse.

- Energy consumption: none (artesian wells and gravity-fed pipes)
- Initial investment: 400 million AMD or US$1 million (including well drilling cost of 6–7 million AMD or US$15,000–17,500 per well)

Water resources assessment (*MCM = million cubic meters)

- Water source: one artesian well
- Permitted amounts: 50 L/s (= 1.6 MCM/yr)
- Actual intake: 200 L/s in 2007 to 150 L/s now (150 L/s = 4.7 MCM/yr) (50 L/s per ha)
- Expected to dry up in three years (the neighboring village is already experiencing negative pressure for their artesian wells and 5 m drop in the shallow groundwater table)
- Drainage: discharged directly to the irrigation drainage right next to the farm

box continues next page

Box 4.4 Sample Fish Farm in the Ararat Valley *(continued)*

Farm budget

- Expenses: 1,000 AMD/kg fish (US$2.5/kg)
- Selling price: 1,800 AMD/kg fish (US$4.5/kg)
- Net profit: 800 AMD/kg fish (US$2/kg)
- Annual Net profit: 50 million AMD or US$120,000/yr (800 AMD/kg or US$2/kg x 60,000 kg)
- Net profit per volume: 12 AMD/m³ or US$0.03/m³ water (50 billion AMD/yr or US$120,000/yr * 1yr/4.7 MCM)
- Productivity: 0.013 kg fish/m³ water (= 60,000 kg fish/4.7 MCM)

Figure 4.2 Discharge of Operating Wells in Ararat Valley in 2007 and 2013

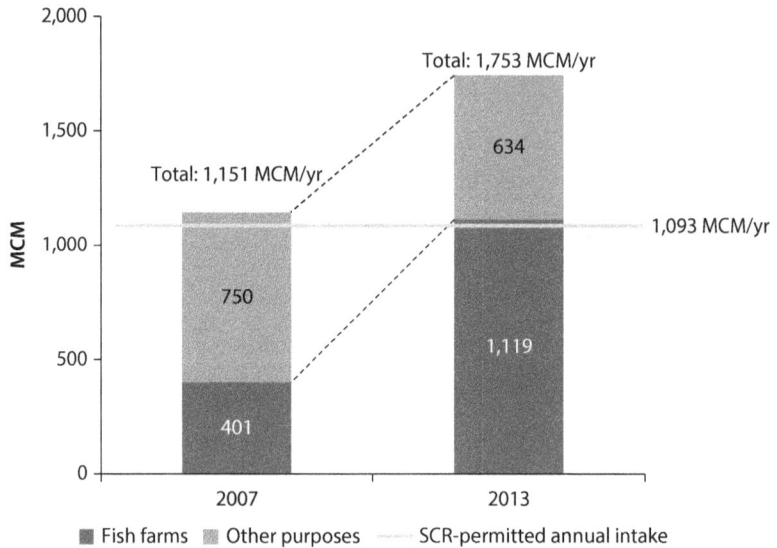

Source: USAID 2014.
Note: "Other purposes" include irrigation, drinking, and industrial water uses; MCM = million cubic meters.

Table 4.8 New Water Use Permits Issued 2008–13

Total number of new permits and abstraction allowed	Of which for:			
	Fishery	*Irrigation*	*Potable*	*Industrial*
274 permits for 735 MCM/yr	202 permits for 700 MCM/yr	56 permits for 32 MCM/yr	12 permits for 2.3 MCM/yr	4 permits for 0.2 MCM/yr

Source: USAID 2014.
Note: MCM = million cubic meters.

water uses (1,337 MCM per year) was less than the amounts of those permits (1,571 MCM per year), total abstraction was greater for two reasons. First, in some cases, fish farms are abstracting more water than allowed in their WUPs. Second, there are illegal wells operating without WUPs. There are 531 wells abstracting 416 MCM per year without WUPs, including 35 fish farming wells discharging 47 MCM per year without WUPs. Table 4.9 summarizes the water uses by fish farms with and without WUPs. Looking at the distribution of fish farms, out of the total 190 in operation, there are 22 fish farms using more than 9.5 MCM per year (or 300 liters per second), and their water uses amount to 57 percent of the total water use of fish farms. Table 4.10 presents the water uses by fish farms taking more than 9.5 MCM per year. Appendix F povides the list of those 22 fish farms and more details on their water uses.

As a result of continued overpermitting and overabstraction of groundwater, the artesian groundwater zone has decreased (map 4.5). Between 1983 and 2013, piezometric levels decreased on average by 6–9 meters, sometimes by as much as 15 meters. Well discharges have reduced by 6–200 liters per second. The artesian zone in the valley has also significantly reduced. The artesian zone decreased from 32,760 hectares in 1983 to 10,706 hectares in 2013. The cone of depression has also extended, now reaching the Sevjur-Aknalich springs, located near Echmiadzin. Flows have decreased to one third of the 2007 levels (from 309

Table 4.9 Fish Farm Wells and Intake in Ararat Valley

No. of fish farms		Number of fish farm wells				Fish farm intake (MCM/yr)		
Total	Actually operating	Total, with permit	Actually operating, with permit	Actually operating, without permit	Wells operating by pumps	Permitted volume	Actual intake, with permit	Actual intake, without permit
267	190	576	470	35	44	1,361	1,072	47 (4.2% of total actual intake, 1,119)

Source: USAID 2013, interim report part 2.
Note: MCM = million cubic meters.

Table 4.10 Water Use by Fish Farms (Water Intake > 300 L/s or 9.5 MCM/yr)

Water intake	Number of operating fish farms	Total water intake by these fish farms	% in total water intake of all fish farms in Ararat valleya	% of water reduction achievable by 70% semi-recycling technology applied to fish farms
> 1,000 L/s (32 MCM/yr)	3	309 MCM/yr	28	19
> 500 L/s (16 MCM/yr)	13	531 MCM/yr	47	33
> 300 L/s (9.5 MCM/yr)	22	639 MCM/yr	57	40

Source: USAID 2013, interim report part 2.
Note: L/s = liters per second; MCM = million cubic meters.
a. Total water intake by all fish farms in Ararat valley is 1,119 MCM/yr.

Map 4.5 Observations on Changes of Groundwater Levels and Artesian Zone in Ararat Valley

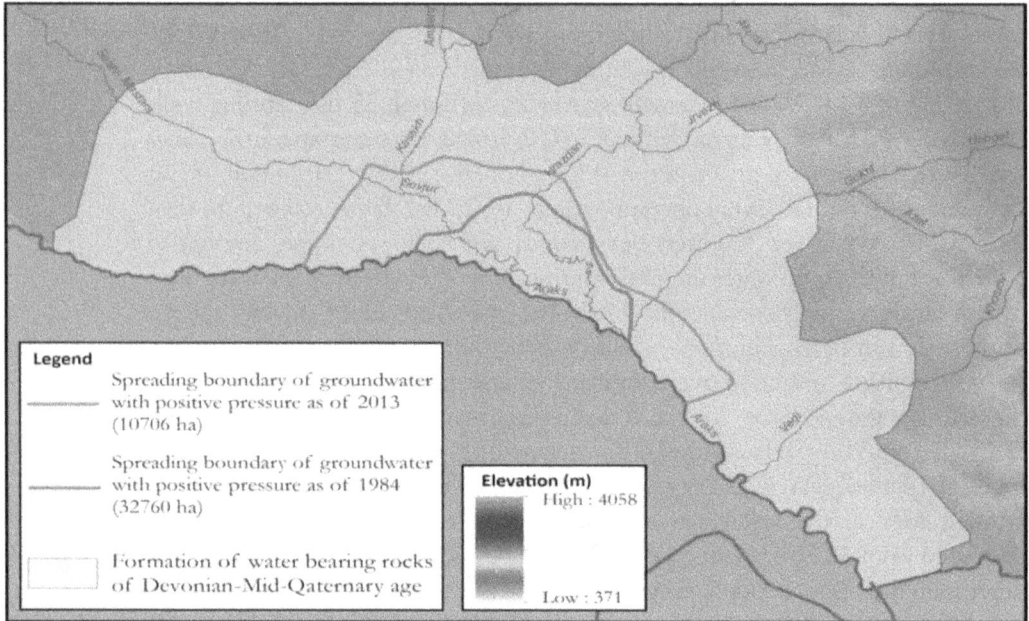

Legend

Spreading boundary of groundwater with positive pressure as of 2013 (10706 ha)

Spreading boundary of groundwater with positive pressure as of 1984 (32760 ha)

Formation of water bearing rocks of Devonian-Mid-Qaternary age

Elevation (m)
High : 4058
Low : 371

Source: USAID 2014.
Note: A full-color version of this map may be viewed at http://www.issuu.com/world.bank.publications/docs/9781464803352.

MCM in 2007 to 95 MCM in 2013). The flow of the Metsamor River, which is fed exclusively from groundwater, has also greatly reduced.

Conflicts with other artesian groundwater users—irrigation, domestic, industrial, and cooling waters—are growing. As the artesian area has reduced in the Ararat valley, the number of communities using artesian wells for irrigation and domestic water supplies has decreased from 44 in 1983 to 13 in 2013. For example, in Echmiadzin, 122 out of 303 previously artesian wells for irrigation and domestic water uses do not flow any more. Due to the reduced discharges of the Sevjur-Aknalich springs, the Armenian (Metsamor) nuclear power plant can take only 16 MCM per year, while its water requirement is 32 MCM per year.

Overloaded Drainage System

Discharge from fish farms is increasing the burden on the agriculture drainage system. There are 1,535 kilometers of operational drainage networks in Ararat valley, including 905 kilometers of open drains and 630 kilometers of closed drains. The drainage system was originally designed for agricultural drainage of up to 1,160 MCM per year. This system is increasingly overloaded with the growth of fish farms (as well as, to a lesser extent, industrial enterprises and communities). In 2012, 1,770 MCM per year was removed by the drainage system (figure 4.3), with about half—877 MCM per year—discharging from fish

Figure 4.3 Annual Discharge of Drainage Network in the Ararat Valley, 1997–2013

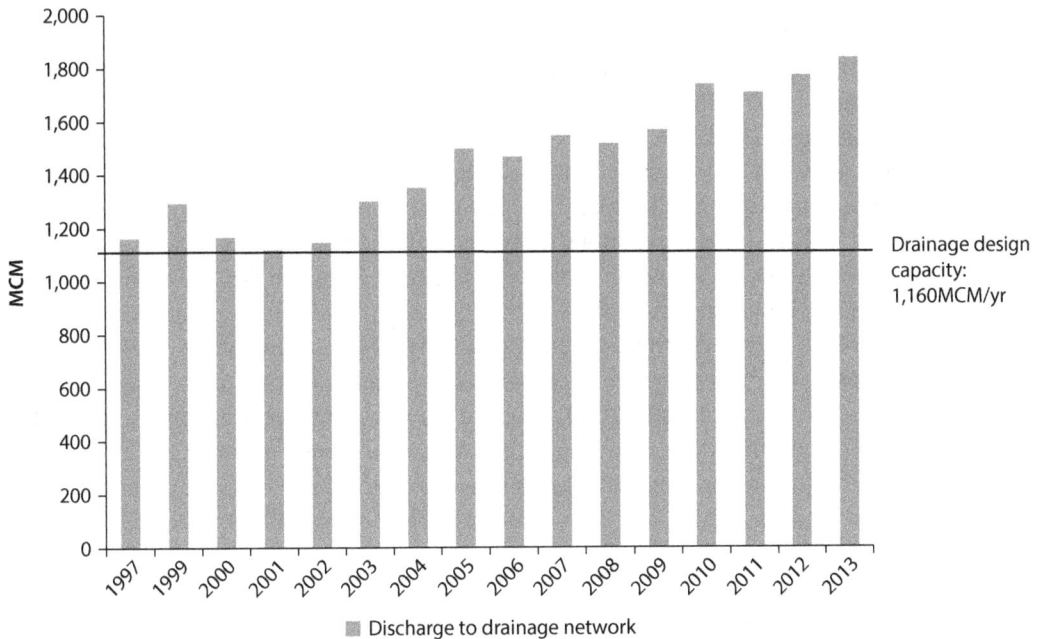

Source: USAID 2013, interim report part 2.
Note: MCM = million cubic meters.

farms in violation of the *water discharge conditions set in their water permits*. This additional discharge results in increases in operation and maintenance requirements (extra financial burden for the State Committee on Water Economy) for the drainage network. Also, the increased water levels in collectors near the drainage network have caused waterlogging, soil salinization, and alkalinization. This ultimately reduces nearby agricultural crop yields. Currently, the State Committee on Water Economy is revising its fees to make fish farms pay 0.33 AMD per cubic meter of water discharged to the drainage system in the Ararat valley.

Efforts to Address the Issue

Recognizing the growing concerns about water resources in the Ararat valley, the government of Armenia established an interagency commission on the issue in late 2010. The commission conducted some research and advised the WRMA to issue no more permits for fish farms, though this was largely not followed. The commission was abolished a year after its establishment due to its ambiguous and weak function. Clearly, there is a need for an effective coordinating mechanism across several departments to monitor status of water resources management issues in the Ararat Valley. In the following years, several measures were adopted by the government, including stricter regulation over water use permitting and enforcement of permit conditions. Restrictions on drilling wells in certain zones of the Ararat valley were also used. As of January 1, 2014, the water abstraction fee was also increased for fish farms. While the water abstraction fee per volume remains the same at 1 Armenian dram per cubic meter, the fee is applied to 50 percent of the overall abstracted volume instead of 5 percent, as was previously done.[7] The Ministry of Territorial Administration is also monitoring water usage by fish farms.

In June 2013, the Ministry of Agriculture instructed fish farms to use semi-closed water recycling (for large, medium, and small fish farms in one, three, and five years, respectively). Many fish farms are opposing such a mandate because of the high cost and perceptions that the quality of the fish will be worse with such technologies. As shown in table 4.10, targeting large fish farms would bring substantial water savings. Several private sector entities and international organizations—for example the Food and Agriculture Organization of the United Nations (FAO)—are providing technical assistance on the use of such recycling technologies. Recently, USAID (2014) assessed the groundwater resources in the valley. The study provides supporting data to assess the current situation and proposes to strengthen monitoring and take appropriate measures (temporary closure, liquidation, and conversion to valve operation) to bring water use to a sustainable level.

Conclusions

Since 2006, there has been uncontrolled development of private fish farms in the Ararat valley, in part due to the rich supply of artesian groundwater of high quality and low cost. The returns to this industry are quite good (boxes 4.3 and 4.4). However, current use by these farms significantly exceeds sustainable yields. Changes in the artesian zone and well pressures are already being observed. Moreover, excessive discharge into the agriculture drainage system is problematic. This is also resulting in conflicts with other artesian groundwater users in the valley—irrigation, domestic, industrial, and cooling water uses. Finally, the situation in the Ararat valley is indicative of the larger problems discussed in the previous chapters with respect to weak monitoring and absence of RBMPs. It is clear that WUPs were issued without a sufficient understanding of the water resources base and the existing water uses in the area. While short-term measures to restore and conserve artesian groundwater are being taken, coordinated action across a variety of departments responsible is urgently needed.

Transboundary Water Resources Issues

Key Messages

- The primary focus of most existing bilateral agreements and treaties between Armenia and its riparian countries, particularly those concluded with the Islamic Republic of Iran and Turkey, relates to water allocation.
- Existing agreements on transboundary waters are silent with regard to groundwater issues.
- Implementation of bilateral agreements between Armenia and Turkey remains low.

- Major water infrastructure plans by Turkey to be used for irrigation, water supply, and hydropower are a major government concern because of the potential impacts.
- To date, not much has been done with respect to bilateral Armenia-Turkey cooperation. The government has expressed willingness to collaborate with Turkey in the construction of a joint multipurpose dam on the Araks River along the Armenia-Turkey border (Surmalu dam) for which a joint technical concept has been prepared.
- The formal role of the WRMA with regard to transboundary water issues is not properly addressed in the current legal framework.
- Lack of formal cooperation between all the riparian countries and lack of legal framework for transboundary cooperation are major limitations for making progress on this front.

Introduction

As indicated in earlier chapters, all of the territory of Armenia is located in transboundary river basins. Important transboundary rivers include the Kura and Araks (map 4.6).[8] The Kura basin is shared with Azerbaijan, Georgia, and Turkey, and the Araks basin is shared with Azerbaijan, the Islamic Republic of Iran, and Turkey. Armenian rivers are tributaries of the Kura and Araks: the Debed, a tributary of the Kura, is shared with Georgia; the Aghstev, also a tributary of the Kura, is shared with Azerbaijan; the Akhuryan, a tributary of the Araks, is shared with Turkey; and the Vorotan, the Arpa, and the Tavush, also tributaries of the Araks, are shared with Azerbaijan. Average annual transboundary surface water inflows and outflows are presented in appendix G. Shared groundwater resources add another level of complexity. Characteristics of principal transboundary aquifers are summarized in table 4.11. According to available information, the Debed aquifer is under the greatest stress (Wada and Heinrich 2013).

Reduction in water availability due to the ongoing developments by Turkey is a major concern for the Armenian government. Existing and planned hydraulic infrastructure in the Araks basin by Turkey for consumptive (irrigation and water supply) and nonconsumptive (hydropower) uses will result in changes in the river flow regime as well as river dynamics and morphology (UNECE 2011).

According to long-term river discharge records of hydrological stations along the Akhuryan and Araks Rivers, which are shown in figure 4.4, a decreasing trend is observed in the flow from the Araks River at the Surmalu station, located downstream of the confluence with the Akhuryan River, even though the Yervandashat station, located upstream of the confluence, shows an increasing trend. As limited information is available on water extractions upstream of the Surmalu station over time, at present it is not possible to categorically conclude that the declining trend in the Araks River is due to upstream extractions for consumptive uses (Hannan, Leummens, and Matthews 2013).

Deterioration of water quality in transboundary rivers is also a concern, for example due to nonpoint source pollution from agriculture and livestock

Map 4.6 Map of Kura-Araks River Basin

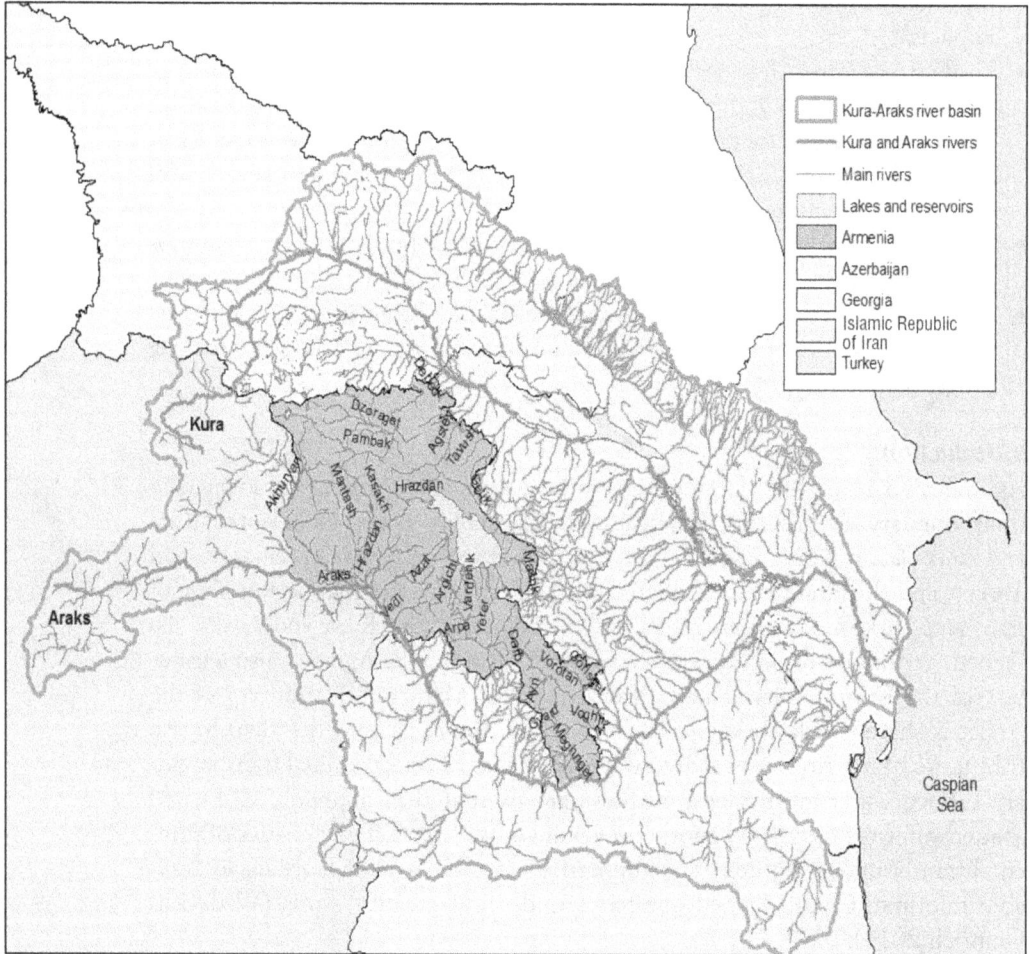

Source: UNDP and GEF 2013.
Note: A full-color version of this map may be viewed at http://www.issuu.com/world.bank.publications/docs/9781464803352.

activities in the Araks and Akhuryan Rivers. Mining is also problematic as it relates to shared aquifers, such as the Aghstev-Tavush and Pambak-Debed aquifers. In these two transboundary aquifers, potential conflicts over the use of readily available resources are also expected as water demand in the riparian countries is increasing (Puri and Aureli 2009).

In addition to transboundary rivers and groundwater, there are important transboundary ecosystems shared by Armenia and Turkey in the Araks/Aras River valley. According to UNECE (2011), the Araks/Aras valley harbors several natural and artificial wetlands that provide important nesting areas for water birds. During the past decade, these wetlands have been under intensive pressure from the increasing development of fish farming. A particularly important site in Armenia is the Khor Virap marsh, which was designated a Ramsar site in 2007.[9]

Past and Ongoing Government Efforts

Development and use of international waters by Armenia is facilitated by a number of bilateral treaties and agreements. Most of them entered into force during the Soviet era. Nonetheless, Armenia has assumed obligations with respect to them. Bilateral agreements on transboundary waters entered into by Armenia and its neighbors are presented in appendix H.

Table 4.11 Characteristics of Principal Transboundary Aquifers

Transboundary aquifer	Countries	Area (ha)	Stress index (low 0, high 1)
Herher, Malishkin, and Jermuk aquifers	Shared between Azerbaijan and Armenia Weak links with surface water	13,066	—
Vorotan-Akora aquifer	Shared between Azerbaijan and Armenia Weak links with surface water	38,771	—
Aghstev/Akstafa-Tavush/Tovuz aquifer	Shared between Armenia and Azerbaijan Groundwater flow from Armenia to Azerbaijan Medium connection with surface water	713,329	0.11
Leninak-Shiraks aquifer	Shared between Armenia and Turkey Groundwater flow from Akhuryan-Arpacay subbasin to Ararat valley Medium links with surface water	516,021	0.04
Debed aquifer	Shared between Armenia and Georgia Alluvial aquifer upper part of the basin and volcanic-sedimentary rocks Medium links with surface water	36,299	0.51

Sources: UNECE 2011; Wada and Heinrich 2013.
Note: — = not available.

Figure 4.4 Time Series Annual Discharge Measured at Upstream and Downstream of the Confluence of the Akhuryan and Araks Rivers

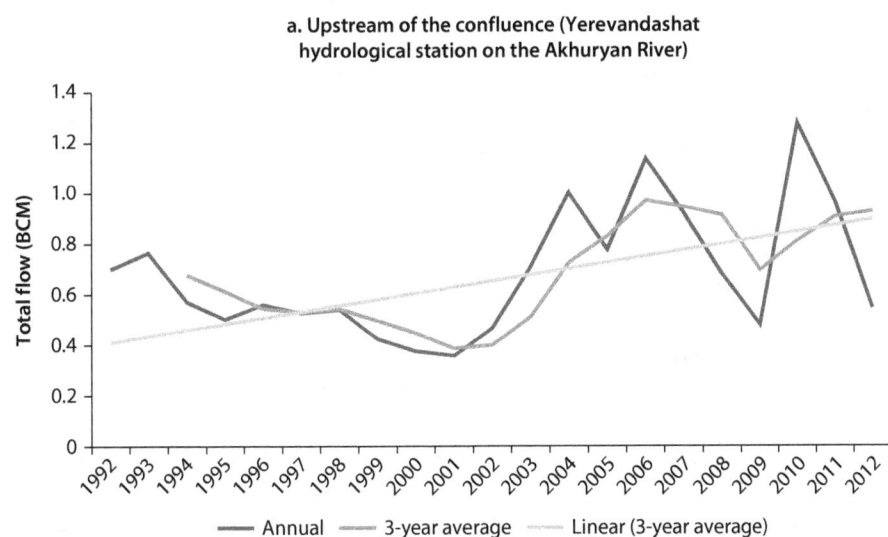

a. Upstream of the confluence (Yerevandashat hydrological station on the Akhuryan River)

figure continues next page

Figure 4.4 Time Series Annual Discharge Measured at Upstream and Downstream of the Confluence of the Akhuryan and Araks Rivers *(continued)*

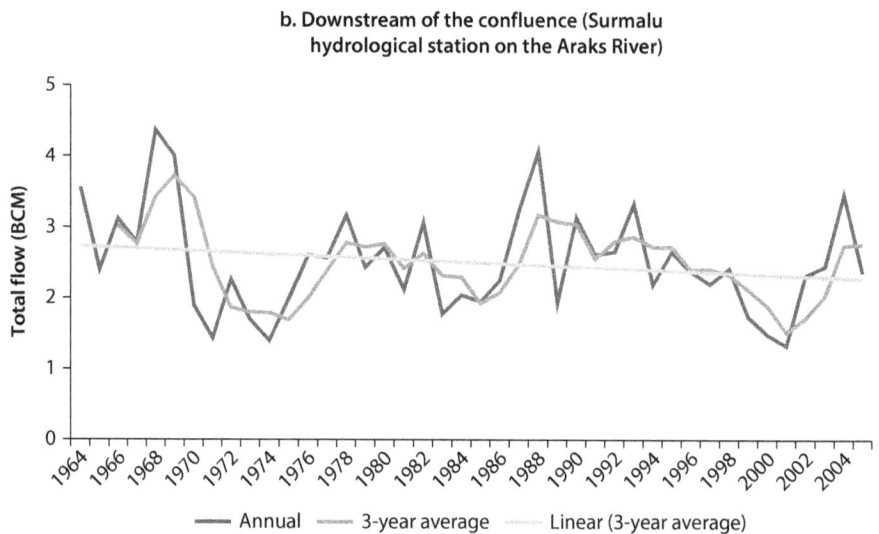

b. Downstream of the confluence (Surmalu hydrological station on the Araks River)

Legend: ■ Annual ■ 3-year average ⋯ Linear (3-year average)

Source: Prepared by authors with data provided by project implementation unit.

Armenia has an agreement with Turkey on the use of the Araks and Akhuryan Rivers. The Kars Protocol, concluded in 1927, includes provisions on the right to use a 50:50 allocation of the flow of the transboundary rivers, small rivers, and streams, as well as several basic regulations on infrastructure and dam construction. A protocol to the above-mentioned agreement was concluded in 1964 on the joint construction of the Akhuryan dam, which provided the basic rules for the joint construction of the dam and the sharing of its water on a 50:50 basis as well as the quantitative regulation of water use downstream of the dam up to the Iranian border. A permanent acting commission was established for the purpose of joint water use and technical exploitation of the Akhuryan reservoir. Another agreement was signed in 1973 on the construction of bridges and border issues on the Akhuryan River, which established basic rules on the regulation of the tributaries. In 1987 a technical and economic report was issued on a proposed reservoir on the Araks River to promote the comprehensive utilization of water resources (for irrigation and power generation) and prevention of channel erosion along the entire length of the Turkish-Armenian border. In 1990 an agreement was signed to address technical issues associated with the construction of joint hydropower facilities, which have not yet materialized, as well as changes in the riverbed and technical cooperation.

An agreement also exists between Armenia and the Islamic Republic of Iran on the joint utilization of the border areas of the Araks River for irrigation, power generation, and domestic use. According to the treaty, the two countries share the waters of the transboundary Araks River on a 50:50 basis. Cooperation schemes were also developed for the construction of joint hydrotechnical facilities, which did not materialize, and the collection of data.

There have also been decrees issued and agreements signed between Armenia and Georgia concerning the use of the Debed River. Similar decrees were passed between Armenia and Azerbaijan concerning the transfer of Arpa River waters into Lake Sevan; the regulation of the Vorotan River flow, which divides the Vorotan flow equally between the two countries and regulates the minimum flow during dry years; and the use of the Aghstev and Tavush Rivers.

Thus, the primary focus of most of the existing bilateral agreements and treaties between Armenia and its riparian countries, particularly those concluded with the Islamic Republic of Iran and Turkey, relates to water allocation. They may need to be revised to take into account water protection considerations. Furthermore, these agreements are silent with regard to groundwater. In general, no detailed provisions related to groundwater are provided. The existing agreements specifically apply to surface water.

Armenia has not signed the 1992 Convention on the Protection and Use of Transboundary Watercourses and International Lakes (Water Convention) concluded under the aegis of UNECE. Concerns associated with the polluter pays principle embodied in the convention have deterred preparation in Armenia. Azerbaijan is the only Armenian neighbor that has ratified the convention. Armenia signed the 1999 Protocol on Water and Health, which is now in the process of ratification. The status of ratification of multilateral treaties and customary international law by Armenia and its neighbors is presented in appendix I.

Although the 2002 Water Code details the protocol to be followed to meet country obligations regarding transboundary waters, including the appointment of permanent inter-State committees for the solution of operational problems, it does not explicitly acknowledge the formal role of the WRMA. The Water Code requires that permanent inter-State committees present their decisions on the operation of transboundary water systems to the State Committee on Water Systems. While this arrangement is appropriate for the management and operation of joint hydrotechnical structures, it seems necessary to assign a specific role to the WRMA to facilitate the broader transboundary cooperation dialogue. The WRMA may play a key role in basic functions such as joint monitoring, data exchange, joint formulation of norms and procedures, identification of investments to optimize the use of joint resources, and negotiation of future agreements (PA Consulting Group 2005).

The Water Code also establishes the Armenian Commission on Transboundary Water Resources. The basic functions of the commission are formulation and submission to the government of draft inter-State agreements, notification to the relevant agencies of issues not regulated by inter-State agreements and requiring due resolution, and the provision of information to agencies in Armenia concerning the state of transboundary waters and transboundary impacts. The chair of the commission is the head of the WRMA, and the members include the deputy chair of the State Committee on Water Economy, the head of the ASHMS, and representatives of the Ministries of Agriculture, Health Care, National Security, and Foreign Affairs; the Water Design Institute; and the Irrigation Water Supply Agency. The commission largely exists on paper only and has no support staff. Thus, transboundary cooperation aspects are dealt with on a case-by-case basis (for example, Armenian-Iranian joint water quality monitoring is coordinated by the WRMA, and Armenian-Turkish joint hydrological measurements are coordinated by the State Committee on Water Economy).

Several donors have supported (and continue to support) transboundary cooperation efforts between Armenia, Azerbaijan, and Georgia. While all these initiatives are considered good attempts to promote cooperation and collaboration to protect transboundary resources, the lack of participation of the Islamic Republic of Iran and Turkey is a major limitation for making major progress on this front. Armenia is conscious of the need for regional engagement between all the riparian countries, in particular with Turkey, and would like to explore cooperation with its neighbor around technical discussions on potential joint investments in the Araks River.

Conclusions

Armenia shares transboundary water resources problems with its neighbors Azerbaijan, Georgia, the Islamic Republic of Iran, and Turkey. While the issues are complex, there is great potential for sharing the benefits of cooperation between the riparian countries in the Kura-Araks basin. The current level of cooperation is weak and ongoing support provided by the donor community to facilitate transboundary cooperation may require further enhancement. A critical area needing consideration includes inviting the Islamic Republic of Iran and Turkey to participate in this dialogue.

Building Water Storage Capacity

Key Messages

- Regulation of surface runoff is of strategic importance for the sustainable development of the irrigation sector in Armenia, particularly in the semiarid regions, where rapidly growing populations are facing depletion of groundwater resources.
- Per capita storage capacity in Armenia is much lower than the capacity of its neighbors, with the exception of the Islamic Republic of Iran.
- A strategic plan for the development of priority reservoirs in Armenia is needed that addresses economic, financial, environmental, and social dimensions, including transboundary impacts, with a sustainable financial approach to provide the needed funding to develop the sites.
- Incomplete dams and existing feasibility studies need to be updated to reassess the technical and economic viability of these investments.
- Large investments should be considered and analyzed within the context of overall river basin planning.

Introduction

Armenian rivers present significant annual and seasonal variability in runoff. In order to address this variability, the country has built 87 dams, with a total capacity of 1.4 billion cubic meters. With the exception of the Marmarik reservoir, which was completed in 2012, all reservoirs were built before and during the Soviet era with the objective of redistributing the river floods on a seasonal or annual basis. As indicated in "Assessment of the water resources baseline" section in chapter 2, most of the reservoirs are considered single purpose, either irrigation or hydropower. The safety conditions of more than 20 of the existing reservoirs, found to pose an imminent threat to human life, were improved under two earlier World Bank-funded projects. Photo 4.3 depicts the Arpilich reservoir.

Photo 4.3 Arpilich Reservoir

Source: Courtesy of Vahagn Tonoyan. Used with permission; further permission required for reuse.

On average, the per capita storage capacity of Armenia is about 450 cubic meters, which is considered low for a semiarid country. In comparison to its neighboring countries (figure 4.5), Armenian per capita storage is similar to that of the Islamic Republic of Iran, and represents less than 20 percent of the storage capacity of Azerbaijan and Turkey and less than 60 percent of the storage capacity in Georgia (FAO Aquastat database).

Recent Government Efforts

Regulation of surface runoff is of strategic importance to the irrigation sector in Armenia. Increasing the strategic water reserves and regulation of river flows is a key action highlighted in the National Water Program. This may be even more critical in the context of future climate change and impacts on the potential

Figure 4.5 Per Capita Storage Capacity in Armenia Compared to Its Neighbors and Other Countries

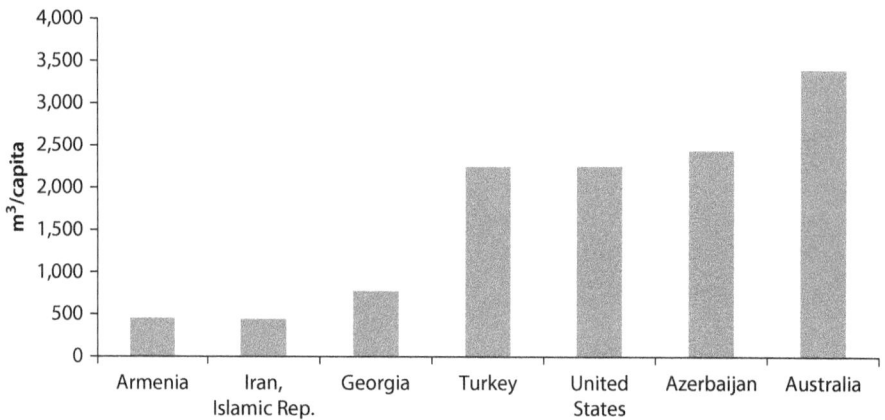

Source: FAO Aquastat database. Based on 2013 data.

frequency and severity of droughts and floods (as noted earlier). According to the Ministry of Territorial Administration, there are 157 potential reservoirs at various stages of construction, design, or planning (table 4.12). Most of the designs were completed during Soviet times. The overall storage capacity of these reservoirs is 1.72 billion cubic meters.

Of the 32 dams either incomplete or at the design stage, 3 are top priorities for the government: Kaps (incomplete), Yegvard (incomplete), and Vedi (at the design stage). In addition, 4 other incomplete reservoirs—the Apna, Karmir Guygh, Artik, and Getik—and 14 new reservoirs—the Lichk, Oshakan, Argichi, Getikvanq, Gegardalich 2, Hartavan, Khndzoreshk, Upper Sasnashen, Elpin, Khachik, Astghhadsor, Byurakan, Geghadzor, and Selav-Mastara—are priorities for the government. Table 4.13 provides a summary of the key features of these priority reservoirs.

During the past few years, the government of Armenia has tried to mobilize external funding for completing the construction of unfinished reservoirs and for updating the feasibility studies of those already designed reservoirs. So far, prefeasibility studies have started for the construction of Vedi reservoir, financed by the French Agency for Development (AFD); and Kaps reservoir, financed by KfW Development Bank (the German development bank). According to the

Table 4.12 Status of Reservoirs in Armenia

Status of reservoirs	Quantity	Storage volume (MCM)
Construction not completed	9	185.4
Designed (different stages of design)	23	733.2
Studied, preliminarily	67	452.8
Planned, but not studied	60	345.9
Total	**157**	**1,717.3**

Source: Water Design Institute of Armenia 2014.
Note: MCM = million cubic meters.

Table 4.13 Key Features of Priority Reservoirs

Reservoir name	River basin	Marz (province)	Status	Total vol. (MCM)	Est. cost[a] (million US$)
Kaps	Akhuryan	Shirak	Partially constructed; feasibility study is in progress for to 60 MCM reservoir option	60.00	44.0
Yegvard	Hrazdan	Kotayk	Partially constructed; feasibility study to be conducted	90.00	139.1
Vedi	Vedi	Ararat	Designed in Soviet times; feasibility study is ongoing; will be followed by preparation of final design for construction of dam	20.00	40.8
Apna	Kasakh	Aragatsotn	Partially constructed; final design was prepared in Soviet times	5.25	8.7
Karmir Guygh	Voskepar	Tavush	Partially constructed	8.50	33.0
Artik	Karkachun	Shirak	Partially constructed	1.69	3.5
Getik	Chichkhan	Lori	Partially constructed; preliminary design available	3.00	7.8
Lichk (Meghriget)	Meghriget	Syunik	New; preliminary design has been prepared by MCA	1.17	6.5
Oshakan (Kasakh)	Kasakh	Aragatsotn	New; feasibility study report is available	13.85	35.0
Argichi	Argichi	Gegharkunik	New dam; preliminary design is available, prepared by Millennium Challenge Corporation	5.50	4.2
Getikvanq	Elegis	Vayots Dzor	New; preliminary investigations have been implemented	23.00	54.0
Gegardalich 2	Yot Aghbyur	Kotayk	New; preliminary design is available	5.50	18.4
Hartavan	Gegharot	Aragatsotn	New; preliminary design is available	3.00	9.7
Khndzoreshk	Karkachun	Syunik	New; preliminary investigations have been implemented	5.20	13.0
Upper Sasnashen	Upper Sasnashen canal	Aragatsotn	New; preliminary investigations have been implemented	1.00	6.5
Elpin	Elpin	Vayots Dzor	New; final design is available	1.00	4.0
Khachik	Khachik canal	Vayots Dzor	New; preliminary investigations have been implemented	0.50	3.1
Astghhadsor	Astghhadsor	Gegharkunik	New; preliminary investigations have been implemented	1.25	2.3
Byurakan (Hamberd)	Hamberd	Aragatsotn	New; preliminary investigations have been implemented	2.70	8.7
Geghadzor	Geghadzor	Aragatsotn	New; preliminary design is available	1.50	6.5
Selav-Mastara	Selav-Mastara	Armavir	New; feasibility study was updated	10.20	32.0
Total				**263.81**	**480.8[b]**

Source: Water Sector Projects Implementation Unit. 2014.
Note: MCM = million cubic meters.
a. Includes design, construction, and technical supervision works.
b. 200 billion AMD.

information provided by the State Committee on Water Economy, the Japan International Cooperation Agency (JICA) and the Kuwait Fund for Arab Economic Development (KF) will likely support the preparation of Yegvard and Selav-Mastara reservoirs respectively. It should be noted that while feasibility studies for these dams are ongoing, technical feasibility has yet to be confirmed. Economic and financial costs and benefits need to be reassessed as well as the integrity of the existing works. Boxes 4.5 and 4.6 provide a few observations from a field visit to the Yegvard and Vedi construction sites.

Box 4.5 Preliminary Assessment of Yegvard Reservoir and Dam

The original dam design was for a height of 48 meters and 228 million cubic meters (MCM) in reservoir capacity. The area to be irrigated from this dam was estimated at 30,500 hectares–7,500 hectares of new irrigated land and 23,000 hectares under improved irrigation. Construction commenced in 1984 but was suspended in 1992. At present, a new proposed dam is contemplated, with a height of 32 meters and 90 MCM in reservoir capacity. The irrigated area is estimated at 11,000 hectares, of which 6,484 hectares of pumping schemes will be converted to gravity irrigation. The estimated cost is around US$87 million or 36 billion (as of 2012) Armenian drams (AMD), including construction and rehabilitation of irrigation canal networks.

According to the new design, the Yegvard dam is expected to store excess water in the Hrazdan River during the winter and supply water for irrigation in the Ararat valley areas, where water levels, pressure, and yields of groundwater in wells and springs have been declining. The Metsamor (Sevjur) River fed from springs also has shown a sharp reduction in available water for irrigation.

The proposed embankment structure design may need to be revised and optimized, considering the anticipated loads, including lower water pressure. Also, detailed geotechnical investigations are needed to assess permeability of the foundation and to propose suitable countermeasures. Currently, more than half of the estimated cost is allocated to antiseepage measures for treating the pervious reservoir floor as per the original design.

The site of the reservoir is located away from the river, and it is to be filled via the Arzni-Shamiran canal. The capacity of this canal is proposed to be increased from 16.6 cubic meters per second to 29 cubic meters per second, possibly with a second feeding canal to fill the reservoir with water from the Hrazdan River during the nonirrigation winter period. Required canal capacity and duration of reservoir filling need to be confirmed.

The potential area benefiting from irrigation is located near the Kasakh River, and a water conveyance system will have to be built. The expected benefits from the dam are increased agriculture productivity, reduction of pumping costs, and reduction of groundwater depletion. The average incremental cost is estimated at US$0.12 or 50 AMD per cubic meter (at 10 percent discount rate).[10]

A feasibility study is needed to assess the economic viability based on topographic, hydrological, geotechnical, and design works, as well as updated agronomic, economic, financial, environmental, and social studies.

Source: Ueda 2012.

Box 4.6 Preliminary Assessment of Vedi Reservoir and Dam

The Vedi dam was designed in 1991 during Soviet times. The original design called for a rock-filled dam of 90.5 meters in height and 38 MCM in reservoir capacity. The estimated area to be irrigated was around 4,000 hectares in the Ararat area. A 21.5-meter-high earth dam would need to be built in the saddle of the reservoir. As the river where the dam is to be built is very dry and seasonal, the original design diverted water from the nearby Vedi River and Khosrov River through an 8.5-kilometer diverting canal and tunnel. The flows to be transferred were estimated at 13 cubic meters per second and 2 cubic meters per second from the Vedi and Khosrov, respectively.

A revised design has been proposed, which includes a smaller dam of 70 meters in height and 20 MCM in reservoir capacity to irrigate 2,745 hectares. This will eliminate the need for a saddle dam and diversion tunnel, requiring only a canal for diversion. The construction cost is estimated at around US$35 million or 15 billion Armenian drams (AMD) (as of 2012). Thus, the average incremental cost is estimated at US$0.22 or 90 AMD per cubic meter (at 10 percent discount rate). The expected benefits from the dam are increased agricultural productivity and reduced pumping costs.

According to information received, the intake site and diversion canal route are still under consideration. Thus, a detailed topographic survey, geotechnical assessment, and updated hydrological assessment are required. The current cost seems to be an underestimate. The cost does not include rehabilitation and construction costs for downstream irrigation networks.

Source: Ueda 2012.

Based on the preliminary construction cost estimates for Yegvard, Kaps, Vedi and Selav-Mastara, and the expected volumes of water to be stored in these reservoirs, the average incremental cost of water ranges between US$0.09 or 37 AMD (Kaps) and US$0.39 or 154 AMD (Selav-Mastara) per cubic meter.[11] At this stage, it seems that the average incremental cost of water is higher than the estimated incremental economic value of water in irrigation.[12] Other economic benefits of the reservoir will have to be identified.

Three key issues that will also need to be considered during the feasibility studies of the priority dams are climate change and transboundary impacts. First, with regard to climate change, as the climate and hydrology have experienced changes since the investments were designed, it is important that the updated feasibility studies include these considerations. As shown in map 4.7, precipitation at the various dam sites has indeed changed over recent decades (on average a decrease in precipitation of about 100 millimeters). Second, with regard to transboundary impacts, as most of the rivers in Armenia are shared with neighboring countries downstream, country impacts would need to be analyzed. Third, these large investments should also be considered and analyzed within the context of overall river basin planning.

Map 4.7 Changes in Precipitation at Dam Sites, 1950–2009

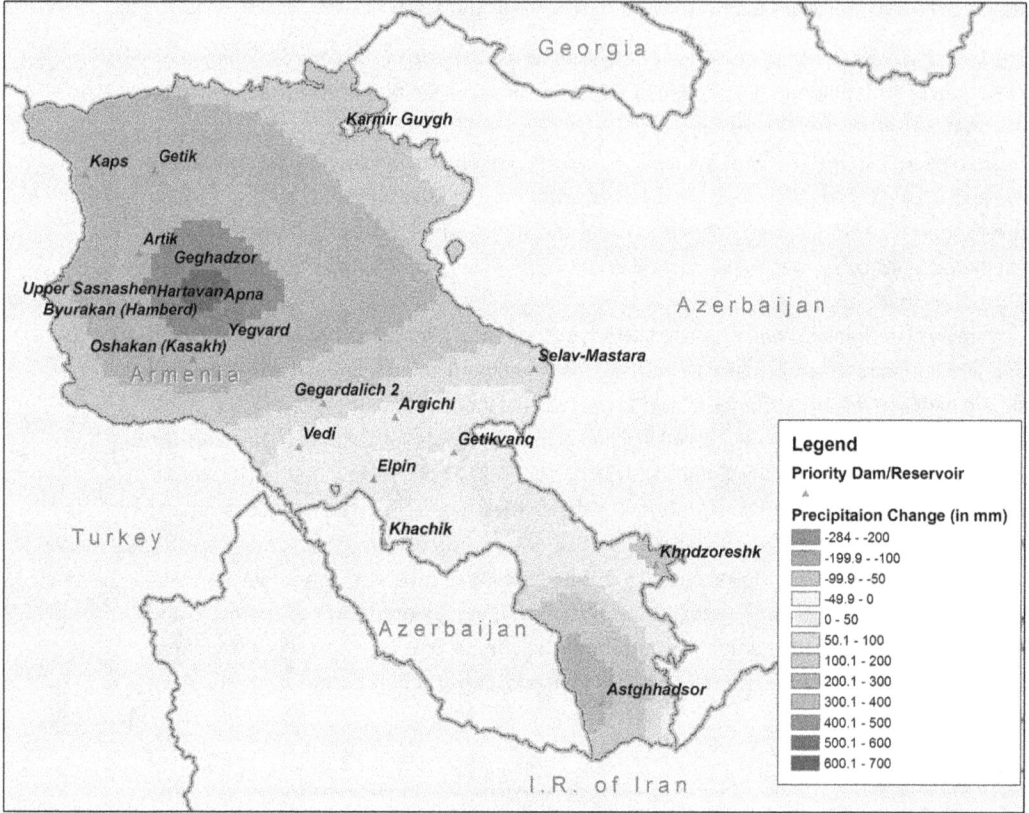

Source: NASA.
Note: Period A is from 1998 to 2009, and Period B from 1950 to 2000. A full-color version of this map may be viewed at http://www.issuu.com
/world.bank.publications/docs/9781464803352.

Conclusions

Storage plays an important strategic role in the regulation of variable surface runoff in Armenia. This is critical for the irrigation, water supply, and energy subsectors, particularly in the semiarid regions where rapidly growing populations are facing depletion of groundwater resources. A comparison with its neighbors shows that Armenia's per capita storage capacity is much lower (with the exception of the Islamic Republic of Iran). Though many of the earlier plans for reservoir development date back to Soviet days, an updated strategic master plan that addresses economic, financial, environmental, and social dimensions, including transboundary and climate change impacts, is missing. Moreover, many of the current incomplete dams and existing feasibility studies would need to be updated to reassess the technical and economic viability of the investments. In addition, an overall financing strategy is needed to support the proposed investments. Finally, as discussed earlier, these large investments should be considered and analyzed within the context of overall river basin planning.

Notes

1. Environmental Impact Monitoring Center: http://www.armmonitoring.am/.
2. ASHMS: http://www.meteo.am.
3. "Tendering of RBMPs for the Akhurian-Metsamor, Chorokhi-Adjaristkali and Upper Kura Basins." Environmental Protection of International River Basins Project: http://blacksea-riverbasins.net/en/tendering-rbmps-akhurian-metsamor-chorokhi-adjaristkali-and-upper-kura-basins.
4. Note that there is an incongruity with respect to mineral water (defined greater than 1 g/l) where the authority over abstraction licenses is with the Ministry of Energy and Natural Resources.
5. There are two artesian aquifers: one of pebble and sand sediments and the other of andesite and basalt rocks. According to investigations conducted in 1958–62, they are located at a depth of 38–180 meters and 25–192 meters, respectively.
6. Currently, the major export markets include Georgia, the Russian Federation, Ukraine, and the United States of America. There is potential to explore the European Union market.
7. The Parliament of Armenia revised the Law on Nature Protection and Nature Utilization Payments, which entered into force on January 1, 2014.
8. Alternative names for the rivers in this section include Kur, Kura (Georgia and Turkey), Mtkvari (Azerbaijan); Araks, Aras (the Islamic Republic of Iran and Turkey), Araz (Azerbaijan); Debed, Dobeda Chay (Georgia); Aghstev, Akstafe (Azerbaijan); Akhuryan, Arpaçay (Turkey); Vorotan, Bargyushad (Azerbaijan); Arpa, Arpa Chay (Azerbaijan).
9. Its importance centers on 100 species of migratory water birds, of which 30 species breed there, including the globally threatened marbled teal and endangered white-headed duck. This site is threatened by a decrease in water level. No management plan has been prepared yet.
10. This is calculated by dividing the present value of all incremental costs (capital, operation, maintenance, and replacement) by the discounted value of the stream of incremental volume of water produced. A 10 percent discount rate is used in the calculation and a 40-year economic life. No physical contingencies are included and the operation and maintenance costs are estimated at 1 percent of construction costs. A four-year construction period for the works is assumed.
11. The calculation assumes a four-year construction period, a 40-year economic life, and a discount rate of 10 percent.
12. Using data from the Implementation Completion and Results Report of the Irrigation Development Project funded by the World Bank, the estimated economic value of water in irrigation ranges between US$0.04 (alfalfa) and US$0.23 (apricots) per cubic meter (World Bank 2009).

References

European Union. 2011. *European Neighborhood Policy Instrument: Shared Environmental Information Systems (ENPI-SEIS), Armenia Country Report.*

FAO (Food and Agriculture Organization of the United Nations). 2011. *Review of Fisheries and Aquaculture Development Potentials in Armenia.*

FAO (Food and Agriculture Organization of the United Nations) Aquastat (database). Rome, Italy. http://www.fao.org/nr/water/aquastat/main/index.stm.

Hannan, T., H. J. L. Leummens, and M. M. Matthews. 2013. *Desk Study: Hydrology*. UNDP/ GEF Reducing Transboundary Degradation in the Kura Araks River Basin Project.

PA Consulting Group. 2005. *Legal and Institutional Reviews of Water Management in Armenia*. Prepared for the USAID Armenia Mission 1.

Puri, S., and A. Aureli. 2009. *Global Atlas of Transboundary Aquifers of the World*. UNESCO IHP Series. Paris: UNESCO Division of Water Sciences. http://www.isarm .org/publications/324.

Ueda, Satoru. 2012. *Armenia Water Resources and Dam Sector Mission Report*. Back-to-Office report prepared after an Armenia Mission in July 24-27, 2012.

UNDP (United Nations Development Programme) and GEF (Global Environment Facility). 2013. "Updated Transboundary Diagnostic Analysis." Prepared for UNDP-GEF Project on Reducing Transboundary Degradation in the Kura Araks River Basin, Baku/Tbilisi/Yerevan, September.

UNECE (United Nations Economic Commission for Europe). 2010. *Policy Brief: Summary of Results and Lessons Learned from the Implementation of the Armenian NPD on IWRM*. Prepared for the National Policy Dialogue in Armenia, November.

———. 2011. *Second Assessment of Transboundary Rivers, Lakes, and Groundwaters*. Convention on the Protection and Use of Transboundary Watercourses and International Lakes.

USAID (United States Agency for International Development). 2005. *Legal and Institutional Reviews of Water Management in Armenia*. Prepared by PA Government Services Inc.

———. 2007. *Compliance Assurance and Enforcement of Water Use Permitting in Armenia: Best Practices and Recommendation for Improvement*. Prepared by PA Government Services Inc.

———. 2008. *Water Resources Atlas of Armenia*.

———. 2012a. *Assessment Study of Ground Water Resources of Ararat Valley: Progress and Next Steps*. Progress report prepared under USAID Clean Energy and Water Program.

———. 2012b. *Conceptual Framework for River Basin Management Planning Process*. Prepared under USAID Clean Energy and Water Program, Program Report No. 5, Mendez England & Associates.

———. 2013. *Analysis and Assessment of Groundwater in Ararat Valley*. Interim reports 1 and 2, prepared under USAID Clean Energy and Water Program.

———. 2014. *Assessment Study of Groundwater Resources of the Ararat Valley*. Final report, prepared under USAID Clean Energy and Water Program.

Wada, Y., and L. Heinrich. 2013. "Assessment of Transboundary Aquifers of the World: Vulnerability Arising from Human Water Use." *Environmental Research Letters* 8: 024003. http://iopscience.iop.org/1748-9326/8/2/024003/pdf/1748-9326_8_2 _024003.pdf.

Water Design Institute of Armenia. 2014. "Concept Paper on Strategy Program of Reservoir Construction in the Republic of Armenia." Water Design Institute of Armenia.

Water Sector Projects Implementation Unit. 2014. *Priority Reservoirs in Armenia*.

World Bank. 2006. *Assessment of Economic Efficiency of Hydrometeorological Services in the Countries of the Caucasus Region*. Internal Interim Report for Discussion.

World Bank. 2009. *Irrigation Development Project: Implementation Completion and Results Report*. Report No. ICR00001145.

CHAPTER 5

Donor Support to the Water Sector in Armenia

Since Armenia became independent from the Soviet Union in 1991, several international financial institutions and bilateral donors have provided technical and financial support to the water sector in Armenia, including the World Bank (WB), International Financial Corporation (IFC), Asian Development Bank (ADB), Eurasian Development Bank (EDB), European Bank for Reconstruction and Development (EBRD), United Nations Development Programme (UNDP), Global Environment Facility (GEF), European Union/European Commission (EU/EC), United Nations Economic Commission for Europe (UNECE), Organisation for Economic Co-operation and Development (OECD), and Organization for Security and Co-operation in Europe (OSCE). The bilateral donors include the United States Agency for International Development (USAID), Millennium Challenge Corporation (MCC), KfW Development Bank (KfW), German Agency for International Cooperation (GIZ), Japanese International Cooperation Agency (JICA), Swedish International Development Cooperation Authority (Sida), Agence Française de Développement (AFD), Kuwait Fund for Arab Economic Development (KF), and the Government of Norway (GoN).

The donors with the most investment and longest history of engagement in the water sector include the WB, EBRD, and USAID. ADB, KfW, and JICA have been active in recent years. AFD and KF are among some of the new donors in Armenia. The areas of engagement by these various donors are given in Table 5.1. More details on the water-related activities supported by the WB and other donors are provided in appendixes J and K, respectively.

Donor Engagement by Topic

- **Monitoring of water quantity and quality.** USAID and the EU have been actively involved in supporting surface water quality monitoring in Armenia. The Environmental Impact Monitoring Center has received modern equipment, including an inductively coupled plasma mass spectrometer, atomic absorption spectrometers, gas chromatographs, and equipment for sampling and analysis of hydrobiological parameters. From 2002 to 2008, USAID also supported the rehabilitation of selected hydrological posts in the Northern and Southern basin management areas, as well as in the Lake Sevan basin and Araks transboundary river. This included installation of continuous stream-gauging devices, acoustic dopplers, and other equipment. In 2008, USAID provided resources to reestablish groundwater monitoring in Armenia. The program assisted in reestablishing the National Reference Groundwater Monitoring Network and provided technical guidelines for groundwater monitoring. The State Water Cadastre Information System (SWCIS), which aims to integrate all water monitoring data from various sources for effective IWRM planning, was also developed with the support of USAID.

- **River basin planning.** To date, several donors have provided financing for the development of river basin plans. This includes financing from the EU for the development of basin plans in the Aghstev, Debed, Akhuryan, and Metsamor (in progress) basins. USAID has supported the drafting of river basin plans in the Vorotan, Meghriget, and Voghji (in progress) basins. UNDP/GEF is financing the preparation of the Arpa basin plan. UNECE supported the preparation of water management measures for the Marmarik basin. None of these plans has yet been formally adopted by the government. Also with support from USAID, a model basin plan was developed to help provide an overarching framework (model contents and analysis).

- **Water use permits.** In 2007, USAID provided a report on international best practices for compliance with the requirements of water use permits, enforcement of the provisions of water use permits, and self-monitoring. The report also made recommendations on how to improve the entire cycle of permit compliance and implementation. In 2008, USAID also supported the development of guidelines for training on compliance as well as procedural aspects of the permitting process.

- **Ararat valley water resources issues.** In 2006–11, the MCC undertook irrigation investments (totaling around US$120 million or 50 billion Armenian drams) to repair gravity-fed irrigation systems and rehabilitate canals, pumping stations, and drainage systems throughout Armenia. In particular, around US$16 million or 6.6 bilion Armenian drams was spent on rehabilitation of some of the drainage infrastructure in the Ararat valley (47.8 kilometers of main drainage canals out of a total 65 kilometers). This work aimed to reduce groundwater levels (from waterlogged lands) in 35 communities of the Ararat and Armavir provinces and to increase crop productivity. As part of the ongoing USAID Clean Energy and Water Program, a comprehensive assessment of the groundwater resources (including artesian aquifers) in the Ararat valley was completed.

- **Transboundary water issues.** There have been several activities by the EU/EC, UNDP/GEF, UNDP/Sida, and OSCE to promote regional dialogue and cooperation on monitoring and management of transboundary water resources. Efforts to date have been mostly focused on the Kura River basin (with Armenia, Azerbaijan, and Georgia). Much of this work has been focused on introducing the principles and approaches of the European Union Water Framework Directive, development of common approaches and methodologies for water quality monitoring and assessment, development of river basin plans in transboundary basins, and the introduction of integrated surface water and groundwater monitoring systems. No donor activity has been successful in engaging the Islamic Republic of Iran and Turkey.

- **New reservoirs.** The government of Armenia has approached several donors for financing for new storage. This remains a long-held priority for the government. Four donors are currently considering four projects: JICA is supporting the feasibility studies for Yegvard reservoir, KfW is supporting the feasibility studies and implementation of Kaps reservoir, AFD is supporting the feasibility studies for Vedi reservoir, and the government of Armenia has requested KfW to undertake preparation for Selav-Mastara reservoir. These four reservoirs were identified during Soviet days as part of the master plan for development of water resources. Most donors agree that there is scope to update this master plan and that it is important to consider new reservoir construction in the context of overall river basin planning.

Donor Engagement in Other Water-Related Issues

Various donors have also provided direct support for water-related issues, such as water supply and wastewater services, hydropower, climate change adaptation, and biodiversity, ecosystem, and environmental protection (including Lake Sevan) (USAID 2012).

Table 5.1 Donor Support to Water Sector in Armenia, by Topic

Category	Challenges and issues	International financial institutions										Bilaterals							
		WB	IFC	ADB	EDB	EBRD	EU/EC	OSCE	UNECE	OECD	UNDP	USAID	MCC	JICA	KfW	GIZ	AFD	KF	GoN
Emerging challenges to IWRM	Future of Ararat valley (agriculture and fishery)																		
	Drainage												X						
	Fishery and groundwater											O							
	Storage and irrigation																		
	New storage													Δ			Δ	Δ	
	Dam safety	X																	
	Irrigation (WUAs, canal rehabilitation, etc)	O			Δ														
	Transboundary issues						X	O			O[a]		X						
	River basin planning (incl. IWRM and institutional/legal reforms)	X					O		O		O	O							
	Monitoring of water quantity and quality																		
	Monitoring capacity											X							X
	Information system											X							
	Issuing, oversight, and control of water use permits											X							
Other water-related issues	Water supply and wastewater services (incl. PPP and water tariff)	O		O		O	O		O	O	X	X			O				
	Energy-water (small, medium, large hydropower plants)	X	X	X		O						O			O				X
	Climate change adaptation (incl. disaster risk management)	X					O	O			O	O							
	Biodiversity, environment, ecosystem protection (incl. Lake Sevan)	X				X	X	X			X					O			
	Solid waste management	Δ		O		O					X								X

Note: X = closed activities; O = ongoing activities; Δ = activities in preparation or prefeasibility assessment; PPP = public-private partnership; WUA = Water User Association.
a. The Reducing Transboundary Degradation of the Kura Aras River Basin Project is implemented by UNDP, with financing from Sida (completed) and GEF (ongoing).

Reference

USAID (United States Agency for International Development). 2012. *Water and Energy Related Donor Projects in Armenia: An Overview Document.* Prepared under USAID Clean Energy and Water Program.

CHAPTER 6

Conclusions and Recommendations

Conclusions

The proper management of water resources will continue to play a key role in the socioeconomic development of Armenia. The performance of the irrigation, hydropower, municipal, industrial, and environment sectors depends on the judicious and wise use of the country's water resources. Though overall water resource availability is good, future pressures (including climate change) may increase tensions across these various water-dependent subsectors. Difficult inter-allocation decisions may need to be made.

Recognizing the importance of integrated water resources management (IWRM), the government of Armenia has introduced over the last decade major institutional and policy reforms. The new Water Code (2002) and subsequent National Water Policy (2005) and National Water Program (2006) provide the legislative foundation and framework (and concomitant institutional bodies and processes) for ensuring the management and development of water resources in the country. Despite this, further institutional strengthening is needed to fulfill the vision of this legislative framework, especially with respect to decen-tralization of roles and responsibilities and the implementation and administra-tion of the water use permit system.

This is now even more important in the context of emerging challenges in the water sector. These challenges includes continued deterioration of the country's monitoring network (both quantity and quality, for both groundwater and surface water), poor water resources planning (from the river basin perspective), continued weak enforcement under the water permit system (the main regula-tory function), concerns over the multitude of water issues in the Ararat valley, increased concerns over transboundary issues, and increased needs for strategic development and management of surface water storage.

Some key conclusions follow.

Obtaining reliable, timely, good-quality, and publicly available data on water quantity and quality are precursors to a properly functioning water management and planning system. The current monitoring system is quite weak and needs substantial investment (both in terms of hardware and human capital).

Monitoring systems are vital to various planning and investment exercises, including in the issuance of and compliance with water use permits. Insufficient investment over decades in the monitoring infrastructure (including institutional capacity building) is evident, and there are opportunities to introduce new technologies and approaches to data collection, verification, and management. Some degree of harmonization across the various departments responsible for monitoring is needed.

River basin management planning needs to be improved, and a strategic vision is required for IWRM in each basin in the country. Despite the various initiatives and multiyear efforts supported by the donor community, the water sector in Armenia still faces many challenges with respect to river basin management planning due to weak capacity and inadequate information and analytical tools. The skills and data needed to carry out modeling and planning work are not yet available within the basin management organizations (BMOs). The current river basin planning model relies heavily on the EU Water Framework Directive and focuses primarily on achieving good ecological status of water bodies. Broad intersectoral planning that takes into account water, agriculture, energy, and environment linkages is not sufficiently developed. No river basin management plan (RBMP) has been finalized and adopted by the government. Government endorsement of such plans is needed to ensure that all levels of government have a consistent approach to water management and clear prioritization of future investments. Nonetheless, the planning of irrigation, water supply, and hydropower investment programs, which are managed at the central level, has limited relationship with the RBMPs. Thus, a clear disconnect exists between the basin plans and sector programs and budgets. Analysis and knowledge of what would be the best allocation (both in terms of economics and efficiency) for the different water users in the basin is needed. Lack of State-level budget is likely to undermine ongoing planning efforts and the full participation of BMOs in river basin planning.

The permitting process is the main regulatory tool for IWRM, but requires more support. The Water Resources Management Agency (WRMA) is the principal agency responsible for issuing water permits. Through decentralization, this function (in the long term) is expected to be devolved to the BMOs. Ensuring compliance with water permits is currently hampered by deficiencies in permitting regulations, insufficient cooperation among agencies, and insufficient resources and weak agency capacity. Currently, compliance involves a monitoring function (WRMA) and an enforcement action function (State Environmental Inspectorate, or SEI). These roles and responsibilities have been separated. Though this separation is advantageous, greater cooperation and coordination (perhaps legislated) on inspection and enforcement is needed between the WRMA and the SEI. Moreover, in the future, compliance history could be made a more explicit part of the permitting process and greater compliance promotion (and more reliance on self-monitoring) undertaken by the government. Refining the permitting procedures for small, medium, and large water uses and pollution discharges may enhance the process, including establishment of a limit of withdrawal and pollution discharge below which a water use permit is not required. Finally, greater public participation in the permitting process may be envisioned to provide greater transparency.

The agriculture and fishery sectors are of strategic importance to the Armenian economy. Ararat valley is the largest agriculture and fish farming zone. Since 2006, a large number of fish farms have been established in the Ararat valley, in part due to the rich supply of high-quality, low-cost artesian groundwater. Due to continued overpermitting and overabstraction, artesian groundwater resources are sharply declining. This is causing conflicts with other water uses in the valley, such as for irrigation and domestic purposes. Several measures are being put in place (for example, adjustments to the abstraction fees). However, coordinated action across several ministries is required.

For Armenia, the transboundary nature of many of the rivers in the country creates a level of water insecurity. Lack of formal cooperation between all the riparian countries and lack of a legal framework for transboundary cooperation are major limitations to making progress on this front. Most of the existing bilateral agreements between Armenia and its riparian countries, particularly those concluded with the Islamic Republic of Iran and Turkey, relate to water allocation. Existing agreements are silent with regard to transboundary groundwater issues. Implementation of bilateral agreements between Armenia and Turkey remains deficient. Major proposed water infrastructure by Turkey (for irrigation, water supply, and hydropower purposes) is a major concern for the government of Armenia because of the expected flow impacts. The government has expressed willingness to collaborate with Turkey on the construction of a joint multipurpose dam on the Araks River along the Armenia-Turkey border (Surmalu dam), for which a joint technical concept has been prepared. Though there are existing arrangements for the management of transboundary waters, the formal role of the WRMA in this regard is not properly addressed in the current legal framework.

Storage plays an important strategic role in the regulation of variable surface runoff in the country. This is critical for the irrigation, water supply, and energy subsectors, particularly in the semiarid regions where rapidly growing populations are facing depletion of groundwater resources. Per capita storage capacity in Armenia is much lower than the capacity of its neighbors, with the exception of the Islamic Republic of Iran. There are a large number of reservoirs that have been in various stages of planning over recent decades. These proposed investments could more than double the existing storage capacity (1.4 billion cubic meters). To move forward, a strategic plan for the development of priority reservoirs in Armenia is needed that addresses economic, financial, environmental, and social dimensions, including transboundary impacts. Many of the earlier master plans were developed during the Soviet era and require updating and revisiting, especially with respect to their current technical and economic viability. This is needed in addition to an overall financing strategy to support the proposed investments. Large investments should also be considered and analyzed within the context of overall river basin planning.

Recommendations

More actions and investment are clearly needed to fully realize the original vision as laid out in the Water Code and subsequent legislation. With the additional pressures and concerns described in the previous chapters, more effort is needed to ensure Armenia's future water security. Table 6.1 synthesizes the recommendations made in this report, and table 6.2 gives some suggested areas where additional financing (and potential additional analytical support) would be required.

Table 6.1 Synthesis of Report Recommendations

Issues	Recommendations
Financial sustainability for IWRM	• Some revision of existing tariff and fee structures may be required • Enhanced budgets to fulfill the mandates of the various institutions given in the existing legislative framework
Weak institutional (capacity) framework	• Continued skills and capacity development of water resource management institutions (particularly water resource management agencies [WRMAs], basin management organizations [BMOs], and water users associations) • Relative responsibilities of various actors need to shift toward greater focus on management
Need for second generation of reforms	• Completion of measures identified in the National Water Program (NWP) • Establishment of Secretariat under the National Water Council (NWC) to monitor and coordinate NWP recommendations and measures
Weak monitoring of water quantity and quality	• Investment in monitoring hardware (both quantity and quality) and staff skills development • Comprehensive review of overall monitoring network and future monitoring needs • Strengthening of public access to water-related data (i.e., revitalize the State Water Cadastre Information System) • Some harmonization across various departments and clarification of roles and responsibilities in monitoring
Weak river basin planning	• Development of skills and capacity (within WRMA and BMOs) for broad river basin planning (with focus on intersectoral concerns and investment planning) • BMOs need to take a more active role during the planning process • Government allocation of budget resources to river basin planning efforts • Government endorsement of existing adequate river basin plans • Enhance economic considerations when preparing river basin management plans
Weak implementation and administration of water permit system	• Governance and transparency issues need to be brought more forcefully • Enhance cooperation among relevant agencies involved with issuance and compliance of permits • Development of skills for compliance assurance • Government allocation of budget resources to the permitting process • Inclusion of compliance history in permitting process • Greater promotion of self-monitoring • Refinement to permitting procedures for different water-use levels • Enhance public participation in the permitting process
Growing water resource concerns in Ararat Valley	• Revisit the water permitting allocations in Ararat Valley • Some further revision of abstraction fees may be needed • Establishment of coordinating mechanism across several departments (e.g., SCWE, Ministry of Agriculture) to monitor status of Ararat Valley • Identification of affordable and economical technologies to reduce water use in fisheries
Growing transboundary water resource concerns	• The formal role for WRMA in transboundary management to be clarified • Revitalize the Armenian Commission on Transboundary Water Resources to more proactively engage in dialogue with its riparian neighbors
Insufficient water storage capacity	• Updating of storage master plans (in the context of river basin plans) to address economic, financial, environmental, and social dimensions • Development of overall financing strategy for proposed storage investments
Donor coordination	• Mechanism needed to coordinate various donors on assistance in the water sector

Table 6.2 Recommendations for Investment and Technical Assistance

Investment and technical assistance requirements	Client	Comments
Strengthening of overall water resources monitoring (including groundwater)	Ministry of Nature Protection	Given the current state of monitoring equipment in the field and the overlapping institutional responsibilities, harmonization and investment is needed. This would include investment in new technologies (both for quality and quantity) and capacity building of various agencies on quality assurance, quality control, data acquisition and storage, etc. Improved groundwater monitoring will be critical. This would support compliance with the European Union Water Framework Directive. A technical audit would be needed to assess the specific requirements, level of investment, and institutional strengthening needed.
Master planning of storage	Ministry of Territorial Administration	Technical assistance is needed to update feasibility studies for the individual reservoirs identified. A larger strategic evaluation and prioritization of all the numerous reservoir proposals is needed. This would look at the full range of economic, financial, environmental, and social issues and provide a framework for future analysis.
Comprehensive development program for Ararat valley	Ministry of Agriculture, Ministry of Nature Protection, Ministry of Territorial Administration	A comprehensive investment project is needed to address the many problems in the Ararat valley. A specific investment in this realm could provide an opportunity (and mechanism) for several ministries to work together. Investments specifically could be in groundwater, fish recycling technologies, drainage improvements, agriculture support, groundwater monitoring, etc.
Institutional strengthening of IWRM	Ministry of Nature Protection	Technical assistance is needed to help build the capacity of the primary IWRM agencies, particularly the WRMA and BMOs. The focus can be on strengthening existing river basin plans, strengthening the water permit process, twinning engagements with international partners on IWRM, etc.

Armenia at a Glance

Table A.1 Armenia at a Glance

Poverty and social	Armenia	Europe & Central Asia	Lower-middle-income	Development diamond[a]
2012				
Population, mid-year *(millions)*	3	271	2,507	
GNI per capita *(Atlas method, US$)*	3,720	6,664	1,893	
GNI *(Atlas method, US$ billions)*	11	1,804	4,745	
Average annual growth, 2006–12				
Population *(%)*	−0.2	0.6	1.5	
Labor force *(%)*	0.9	1.3	1.4	
Most recent estimate *(latest year available, 2006–12)*				
Poverty *(% of population below national poverty line)*	36	–	–	
Urban population *(% of total population)*	64	60	39	
Life expectancy at birth *(years)*	74	72	66	
Infant mortality *(per 1,000 live births)*	15	19	46	
Child malnutrition *(% of children under 5)*	5	2	24	
Access to an improved water source *(% of population)*	100	95	88	
Literacy *(% of population age 15+)*	100	98	71	
Gross primary enrollment *(% of school-age population)*	102	101	106	
Male	96	101	107	
Female	110	100	104	

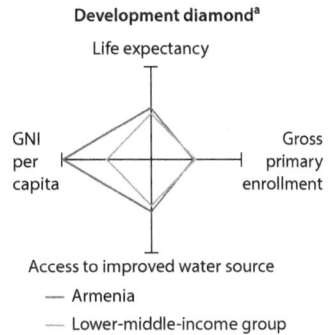

Development diamond[a]: Life expectancy; GNI per capita; Gross primary enrollment; Access to improved water source. — Armenia — Lower-middle-income group

table continues next page

Table A.1 Armenia at a Glance (continued)

Key economic ratios and long-term trends	1992	2002	2011	2012
GDP (US$ billions)	1.3	2.4	10.1	10.0
Gross capital formation/GDP	1.6	21.7	27.3	23.8
Exports of goods and services/GDP	39.8	29.4	23.8	25.1
Gross domestic savings/GDP	−19.8	4.4	3.7	−0.4
Gross national savings/GDP	–	15.4	17.2	10.8
Current account balance/GDP	−13.4	−6.2	−12.4	−11.9
Interest payments/GDP	–	1.3	1.7	1.6
Total debt/GDP	–	72.0	72.8	76.5
Total debt service/exports	–	9.9	27.6	32.3
Present value of debt/GDP	–	–	–	59.2
Present value of debt/exports	–	–	–	163.7

Economic ratio[a]

Trade

Domestic savings — Capital

Indebtedness

— Armenia
— Lower-middle-income group

	1992–02	2002–12	2011	2012	2012–16
(averge annual growth)					
GDP	5.4	6.2	4.7	7.2	–
GDP per capita	6.6	6.6	4.7	7.0	–
Exports of goods and services	−3.2	2.0	14.7	10.7	5.8

Structure of the economy	1992	2002	2011	2012
(% of GDP)				
Agriculture	31.0	26.0	22.7	21.6
Industry	39.4	39.0	33.1	33.2
Manufacturing	33.1	16.7	11.2	11.2
Services	29.6	35.1	44.2	45.2
Household final consumption expenditure	101.3	85.6	83.4	87.5
General gov't final consumption expenditure	18.5	10.0	12.9	13.0
Imports of goods and services	61.3	46.6	47.4	49.3

Growth of capital and GDP (%)

— GCF — GDP

	1992–02	2002–12	2011	2012
(average annual growth)				
Agriculture	2.9	4.7	14.0	9.5
Industry	6.5	4.1	0.0	5.2
Manufacturing	4.3	4.3	12.4	3.1
Services	6.1	8.4	3.4	6.5
Household final consumption expenditure	2.8	6.0	5.3	4.1
General gov't final consumption expenditure	−0.2	6.1	1.9	0.2
Gross capital formation	15.6	7.1	−13.3	−2.0
Imports of goods and services	−2.5	3.8	−1.4	−3.0

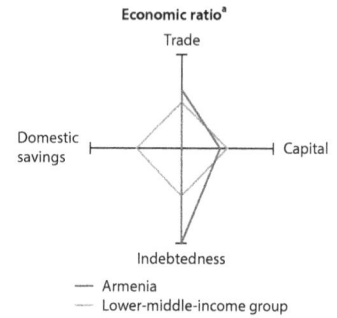

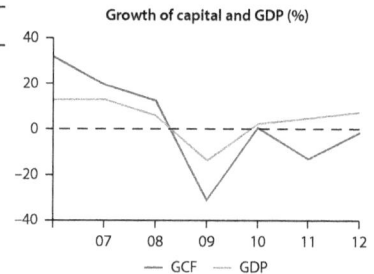

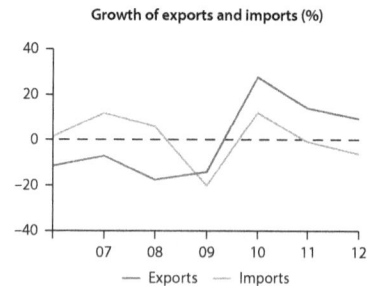

Growth of exports and imports (%)

— Exports — Imports

table continues next page

Table A.1 **Armenia at a Glance** (continued)

Prices the government finances	1992	2002	2011	2012
Domestic prices				
(% Change)				
Consumer prices	–	1.1	7.7	2.6
Implicit GDP deflator	568.8	2.4	4.3	−1.3
Government finance				
(% of GDP, includes current grants)				
Current revenue	4.0	16.6	23.3	22.5
Current budget balance	−7.7	0.5	−0.3	−1.5
Overall surplus/deficit	−7.7	−2.6	−6.5	−7.9

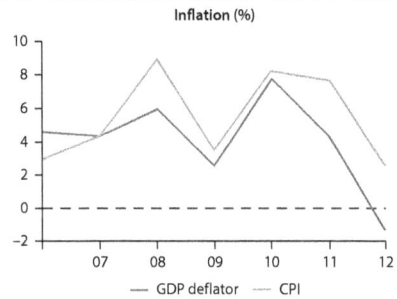
Inflation (%) — GDP deflator, CPI

Trade	1992	2002	2011	2012
(US$ millions)				
Total exports (fob)	220	505	1,284	1,393
Gold, jewelry, and other precious stones	–	258	179	172
Machinery and mechanical equipment	–	41	46	41
Manufactures	–	81	197	209
Total imports (cif)	334	987	4,207	4,208
Food	–	200	799	747
Fuel and energy	60	171	784	801
Capital goods	–	191	1,188	1,212
Export price index (2000 = 100)	–	99	133	132
Import price index (2000 = 100)	–	99	186	170
Terms of trade (2000 = 100)	–	100	72	78

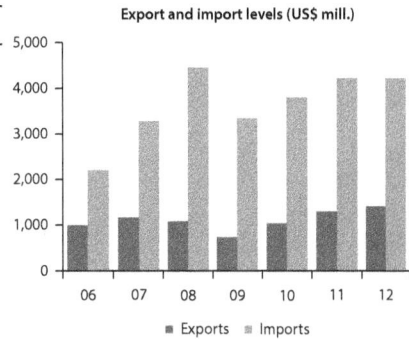
Export and import levels (US$ mill.) — Exports, Imports

Balance of payments	1992	2002	2011	2012
(US$ millions)				
Exports of goods and services	230	698	2,284	2,409
Imports of goods and services	364	1,107	4,679	4,699
Resource balance	−135	−409	−2,395	−2,290
Net income	−39	88	559	629
Net current transfers	–	173	818	638
Current account balance	−171	−148	−1,261	−1,183
Financing items (net)	–	229	1,415	1,271
Changing in net reserves	–	−81	−154	−87
Memo:				
Reserves including gold (US$ millions)	1	431	1,932	1,799
Conversion rate (DEC, local/US$)	0.2	573.4	372.5	401.8

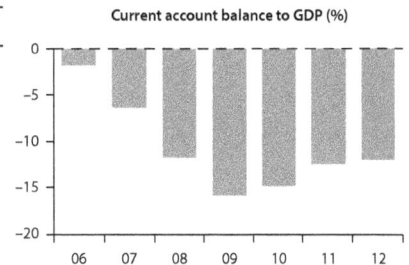
Current account balance to GDP (%)

table continues next page

Toward Integrated Water Resources Management in Armenia • http://dx.doi.org/10.1596/978-1-4648-0335-2

Table A.1 Armenia at a Glance (continued)

External debt and resource flows	1992	2002	2011	2012
(US$ millions)				
Total debt outstanding and disbursed	–	1,712	7,383	7,608
IBRD	–	8	151	236
IDA	–	530	1,187	1,235
Total debt service	–	83	926	1,159
IBRD	–	1	2	3
IDA	–	3	27	27
Composition of net resource flows				
Official grants	19	77	226	137
Official creditors	–	64	175	240
Private creditors	–	–6	578	544
Foreign direct investment (net inflows)	2	111	663	489
Portfolio equity (net inflows)	0	0	0	2
World Bank program				
Commitments	–	9	141	118
Disbursements	–	66	95	151
Principal repayments	–	0	19	19
Net flows	–	66	76	132
Interest payments	–	4	10	11
Net transfers	–	62	66	121

Composition of 2012 debt (US$ mill.)

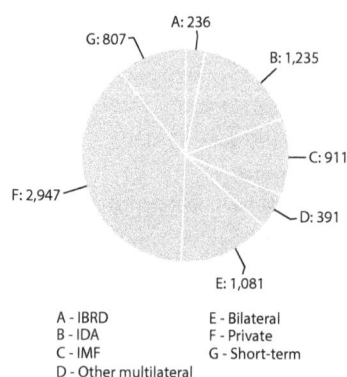

A: 236
B: 1,235
C: 911
D: 391
E: 1,081
F: 2,947
G: 807

A - IBRD   E - Bilateral
B - IDA   F - Private
C - IMF   G - Short-term
D - Other multilateral

Source: World Bank Development Economics LDB database.
Note: — = no data available.
a. The diamonds show four key indicators in the country (in bold) compared with its income-group average. If data are missing, the diamond will be incomplete.

Basic Water Balance

Table B.1 Discrepancies in Water Balance Components

| | Climate | | Surface water (SW) | | | | Groundwater (GW) | | | | | | Total | |
	Precipitation	Evaporation	River flow	Transboundary river flow (Armenia portion)	Depth inflow	Total SW resources	GW Spring flow	GW Drainage flow	GW Deep flow	GW inflow (NWP)	GW outflow (NWP)	Total GW resources	Total renewable water	Usable water resource
Republican Geological Fund (1984)	—	—	—	—	—	—	—	—	—	1.132	0.400	GW produced internally: 4.217 Total GW: 4.950	—	—
National Water Policy (Rep. of Armenia 2005)	17.600	11.475	6.250	0.94	—	Total renewable SW: 7.190 (= 6.250 + 0.94, excluding Lake Sevan)	1.595	1.434	—	1.193	1.068	4.017	—	—
National Water Program (Rep. of Armenia 2006)	—	—	6.859	1.190	—	Total usable SW: 8.049	—	—	—	—	—	GW: 3.611 Renewable GW: 1.000	9.049	(= 8.049 + 1.000)
Water Atlas (USAID 2008)	18.760	10.832	6.775	—	0.611	—	1.594	1.434	0.989	—	—	4.017 (1.594 + 1.434 + 0.989)	—	9.049
FAO Aquastat (2011 data)	16.71	—	—	0.91 (accounted flow of border rivers)	—	SW produced internally: 3.948 SW leaving the country not subject to treaties: 5.28 Total renewable SW: 4.858 (= 0.91 + 3.948 − 5.28)	—	Overlap between SW and GW: 1.4	—	—	—	GW produced internally: 4.311 Total renewable GW: 4.311	7.769 (= 4.858 + 4.311 − 1.4)	—

Source: Republic of Armenia 2005. Republic of Armenia 2006. USAID 2008. FAO Aquastat database (2011 data), Republican Geological Fund 1984.

Note: All numbers are in billion cubic meters per year. — = no data available. *Spring flow* is artesian groundwater discharge. These values are based on field hydrogeological studies. *Drainage flow* is base flow from shallow groundwater aquifers and is based on measurements in different river sections when there has been no precipitation. *Deep flow* is calculated from the water balance.

References

FAO (Food and Agriculture Organization of the United Nations) Aquastat (database). Rome, Italy. http://www.fao.org/nr/water/aquastat/main/index.stm.

Republic of Armenia. 2005. *Republic of Armenia Law on Fundamental Provisions of the National Water Policy.*

———. 2006. *Republic of Armenia Law on National Water Program.*

USAID (United States Agency for International Development). 2008. *Water Resources Atlas of Armenia.*

Republican Geological Fund. 1984. *Assessment of fresh ground water resources of the Armenia Soviet Socialistic Republic.* Note: Republican Geological Fund is a part of the Geological Agency, Ministry of Energy and Natural Resources.

Differences between Global Climate Models on Change in Annual Precipitation and Temperature by the 2050s

Figure C.1 Change in Annual Precipitation by the 2050s

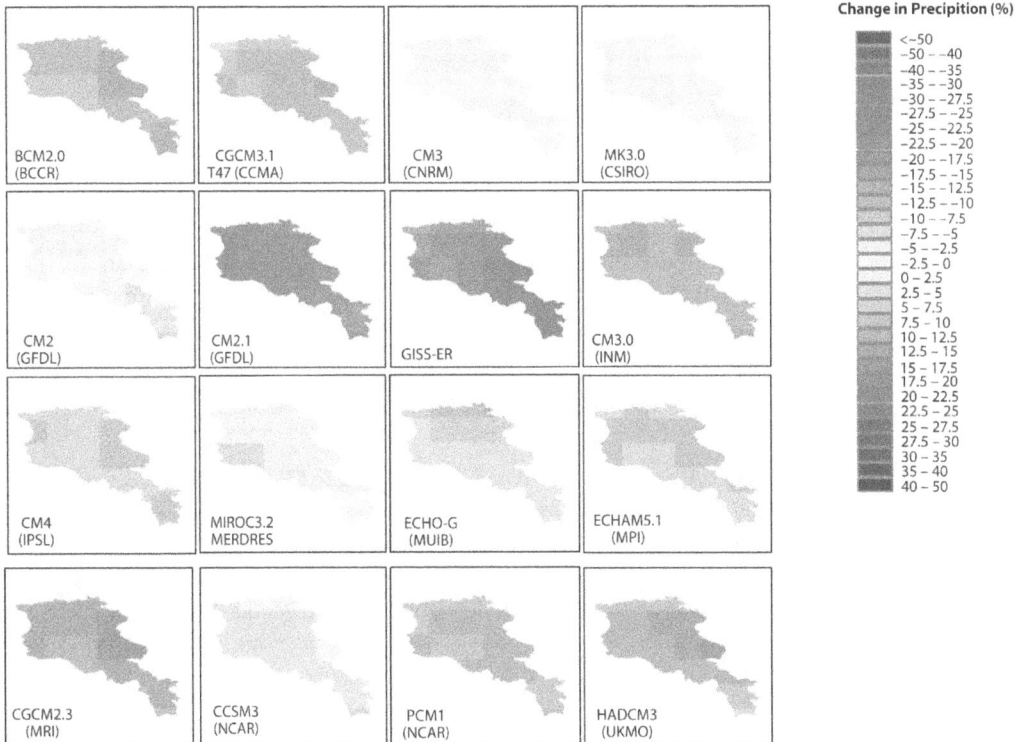

Change in Precipitation (%)

- <−50
- −50 – −40
- −40 – −35
- −35 – −30
- −30 – −27.5
- −27.5 – −25
- −25 – −22.5
- −22.5 – −20
- −20 – −17.5
- −17.5 – −15
- −15 – −12.5
- −12.5 – −10
- −10 – −7.5
- −7.5 – −5
- −5 – −2.5
- −2.5 – 0
- 0 – 2.5
- 2.5 – 5
- 5 – 7.5
- 7.5 – 10
- 10 – 12.5
- 12.5 – 15
- 15 – 17.5
- 17.5 – 20
- 20 – 22.5
- 22.5 – 25
- 25 – 27.5
- 27.5 – 30
- 30 – 35
- 35 – 40
- 40 – 50

Sources: Meehl et al. 2007 (WCRP's CMIP3), downscaled by Maurer, Adam, and Wood 2008.
Note: This figure shows the precipitation change projected by the considered climate model, under the A2 scenario for 2040–2069 as compared with 1961–1990. Map displays gridded data (cell size = 0.5dd). A full-color version of this figure may be viewed at http://www.issuu.com/world .bank.publications/docs/9781464803352.

Figure C.2 Change in Annual Temperature by the 2050s

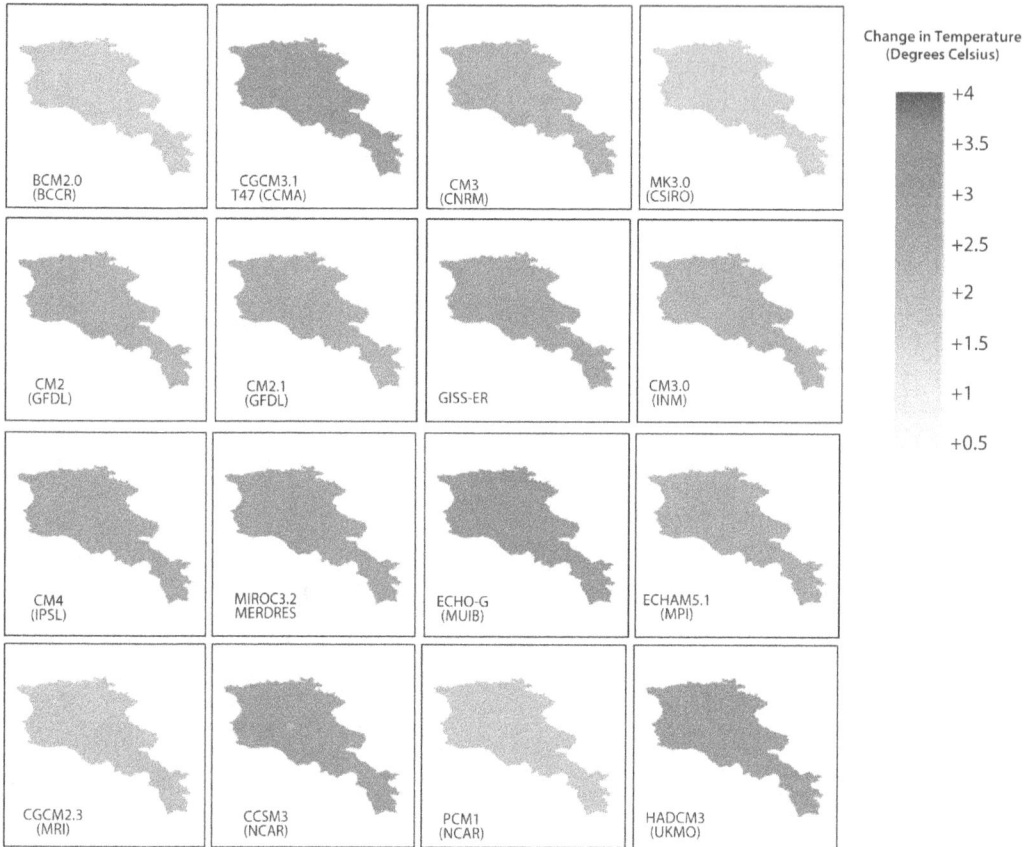

Sources: Meehl et al. 2007 (WCRP's CMIP3), downscaled by Maurer, Adam, and Wood 2008.
Note: This figure shows the temperature change projected by the considered climate model, under the A2 scenario for 2040–69 as compared with 1961–90. Map displays gridded data (cell size = 0.5dd). A full-color version of this figure may be viewed at http://www.issuu.com/world.bank .publications/docs/9781464803352.

References

Maurer, E. P., J. C. Adam, and A. W. Wood. 2009. "Climate Model Based Consensus on the Hydrologic Impacts of Climate Change to the Rio Lempa Basin of Central America." *Hydrology and Earth System Sciences* 13: 183–94.

Meehl, G. A., C. Covey, T. Delworth, M. Latif, B. McAvaney, J. F. B. Mitchell, R. J. Stouffer, and K. E. Taylor. 2007. "The WCRP CMIP3 Multi-Model Dataset: A New Era in Climate Change Research." *Bulletin of the American Meteorological Society* 88: 1383–94.

Status of Implementation of the National Water Program

Table D.1 Status of Implementation of the Short-Term Measures of the National Water Program

Issue	Short-term measure	Implementation status	Challenges and opportunities
Legal requirements			
Harmonization, completion, and improvement of the legislative basis	1. Intersectoral harmonization and improvement of the existing legislation, including analysis of application of the requirements of article 121 of the Water Code and implementation of functions stipulated thereby (main authorities responsible).	The requirements of article 121 of the Water Code are mainly implemented, and parallel to that periodic harmonization of intersectoral legislation is being implemented.	Despite significant progress in this direction, and development of over 120 bylaws to ensure the smooth implementation of the Water Code, National Water Policy, and National Water Program, there are some discrepancies between certain legal acts (e.g., between the Water Code and the Law on Groundwater). There is a need to establish clearinghouse mechanisms to help work toward harmonization and improvement of the legislative basis. The newly established Water Resources Policy Division of the Ministry of Nature Protection could play this role.
	2. Establishment of an interagency standing commission within the National Water Council for ensuring the discussion of changes and amendments to be made to the legal acts.	Not completed.	Though such an interagency standing commission has not been established, under the umbrella of the European Union Water Initiative a steering committee of the National Policy Dialogue on IWRM in Armenia has been successfully operating since 2007. The steering committee, which includes representatives of the Ministries of Nature Protection, Agriculture, Health Care, Energy and Natural Resources, Finance, Emergency Situations, and Territorial Administration, as well as representatives of academic and higher educational institutions, has already had 11 meetings since its establishment. Water policy issues, including legislative issues, are being discussed in the steering committee meeting. If the official status of the group is formally changed, it can quite well serve as the interagency standing commission.

table continues next page

Table D.1 Status of Implementation of the Short-Term Measures of the National Water Program *(continued)*

Issue	Short-term measure	Implementation status	Challenges and opportunities
Institutional development			
Clarification of roles and responsibilities of the water sector institutions	3. Review and implementation of developed recommendations related to overlaps and gaps in the roles and responsibilities identified during the institutional and legal assessments.	The last comprehensive legal and institutional review of water management in Armenia was conducted in 2005 by the USAID Program for Institutional and Regulatory Strengthening of Water Management in Armenia. The assessment also includes recommendations and a proposed action plan to implement the recommendations. However, most of the recommendations and proposed actions have not been implemented so far.	In recent years in both the institutional and legal framework of water management in Armenia significant changes have occurred (such as establishment of a new Water Policy Division in the Ministry of Nature Protection or development of new surface water quality norms). Taking this into account, it is necessary to conduct a new and comprehensive assessment of the legal and institutional framework to clarify the roles and responsibilities of corresponding institutions, and develop corresponding roadmaps for development of those institutions.
Improvement of interagency cooperation and coordination	4. Adjustment and improvement of the mechanisms for interagency cooperation and coordination by the National Water Council.	Not completed. Despite the fact that recently the National Water Council has been having more frequent meetings, so far no major improvements have been made in interagency cooperation.	The expected result of this activity is more open communication among stakeholders and greater data and information exchange. Under the umbrella of the European Union Water Initiative, a steering committee of the National Policy Dialogue on IWRM in Armenia has been successfully operating since 2007. The steering committee, which includes representatives of the Ministry of Nature Protection, Agriculture, Health Care, Energy and Natural Resources, Finance, Emergency Situations, and Territorial Administration, as well as representatives of academic and higher educational institutions, has all the capacities to serve as the main mechanism of interagency cooperation and coordination.

table continues next page

Table D.1 Status of Implementation of the Short-Term Measures of the National Water Program *(continued)*

Issue	Short-term measure	Implementation status	Challenges and opportunities
Development of the basin management organizations (BMOs)	5. Development of a program for institutional development of the BMOs.	Not completed. Though recently the WRMA started to develop annual workplans for BMOs, those workplans mainly relate to implementation of their everyday function, and do not contain an institutional development component.	Unless BMOs have more authority in water resources management in their respective basins, it will be challenging to develop and implement an institutional strengthening program. Despite the objective of the Ministry of Nature Protection to support decentralized management of water resources, BMOs still do not have enough authority to issue permits for water resources of even local importance.
Water resources management needs			
Development of new programs for monitoring of surface water and groundwater resources	6. Development and testing of a pilot monitoring system in one basin management area.	Within the framework of the medium-term expenditure framework, State budget funding was provided in 2009 to implement a project that will serve as a model for replication in other river basins.	The Ministry of Nature Protection is in the process of establishing an interagency working group on water quality monitoring and water quality assessment, which in the next two to three years should work on transition toward a Water Framework Directive–compliant water quality monitoring and assessment system. The working group should carefully review the monitoring systems developed and tested in the pilot basins of Debed, Aghstev, Akhuryan, and Metsamor (Sevjur), and explore the option of replicating them throughout the country.

table continues next page

126

Table D.1 Status of Implementation of the Short-Term Measures of the National Water Program *(continued)*

Issue	Short-term measure	Implementation status	Challenges and opportunities
	7. Development of a monitoring strategy and a national program, technical capacity building of the regional subdivisions of monitoring services, needs assessment, structural improvements and technical modernization, and establishment of an electronic data exchange system.	With the support of the European Union Kura River Phase II project, a monitoring system compliant with the European Union Water Framework Directive has been proposed for the Aghstev and Debed River basins of Armenia, which includes biological, hydromorphological, and physical-chemical monitoring. The proposed monitoring has been successfully tested in the Debed River basin with the support of the European Union Kura River Phase III Project (in 2012) and the European Union Environmental Protection of International River Basins Project. However, this does not include groundwater resources. In 2014 the Environmental Protection of International River Basins Project will propose and test a Water Framework Directive-compliant surface water and groundwater monitoring system in the Akhuryan and Metsamor (Sevjur) River basins of Armenia, which afterwards can be replicated throughout the country.

The project application submitted in the framework of the medium-term expenditure framework has not been approved due to absence of funding. | The interagency working group on water quality monitoring and assessment, which is in the process of formation and which will be working on development of a water quality monitoring and water quality assessment system in the next two to three years, can in parallel work on development of a monitoring strategy, as the two issues are interrelated. |

table continues next page

Table D.1 **Status of Implementation of the Short-Term Measures of the National Water Program** *(continued)*

Issue	Short-term measure	Implementation status	Challenges and opportunities
	8. Reestablishment of the groundwater resources monitoring system in Armenia.	According to government Decree No. 1616-N of September 8, 2005, the Hydrogeological Monitoring Center was established as a State noncommercial organization under the Ministry of Nature Protection. It aims to evaluate the main patterns of formation of freshwater underground waters in the territory of Armenia, and their quantitative and qualitative properties and regional changes, and use this information for more efficient use and protection of groundwater resources of the country, and for the development of measures to combat negative impacts on groundwater resources. In 2006–08 the monitoring of groundwater resources conducted by the Hydrogeological Monitoring Center was fragmented due to insufficient financial resources, but since 2009 the monitoring network has been extended and consists of 70 observations wells and springs. The monitoring includes measurements of water spring and water discharge levels (pressure) and water temperature.	Still, the national reference hydrogeological monitoring network does not cover all the river basins of Armenia. According to experts at the Hydrogeological Monitoring Center, the number of monitoring stations should be about three times more to cover the entire territory of the country (as it used to in the 1980s). However, due to difficulties in funding, the Hydrogeological Monitoring Center did not get any budgetary increase to expand its monitoring network. Solution to this could be installation of automatic monitoring of water levels (piezometers) that would be a one-time cost but would significantly reduce travel costs and facilitate high-quality monitoring and data storage.
Improvement of water use permit procedure	9. Improvement of the existing water use permit regulations, and establishment of criteria for assessment of priority of water use application.	A project has been implemented within the framework of State budget funding in 2009. Parallel to that, the government has approved Resolution 677-N of May 12, 2011, on making changes to Resolution 218-N of March 7, 2003, on an Exemplary Form of Water Use Permit and Approving Water Use Permit Forms, the application of which promotes transparent and informed decision making and provides for establishment of an efficient and easy-to-apply system for water use permits.	One of the main criteria that could greatly help decision makers in assessing the priority of water use applications is cost-benefit analysis of the use of 1 cubic meter of water within the same water use sector (if water use sectors are different, then priorities are set in the National Water Policy in the following order: domestic; agricultural; energy; industrial; recreational; local, regional, and national development; drought control).

table continues next page

Table D.1 Status of Implementation of the Short-Term Measures of the National Water Program *(continued)*

Issue	Short-term measure	Implementation status	Challenges and opportunities
	10. Development of criteria and guidelines for environmental impact assessment as part of the water use permit application process, in cases when possible substantial impact on the environment can occur from a given water use.	A project was implemented in 2009 within the framework of State budget funding. Parallel to that, the government has developed and approved the following decisions: 1. Resolution 927-N of June 30, 2011, on Defining Drinking-Household and Agricultural Water Demand and Assessing the Environmental Flow According to River Basin Management Area. 2. Resolution 118-N of January 14, 2010, on Defining Measures for Application of Modern Technologies, Improving Water Resources Monitoring, and Reducing and Preventing Pollution.	These decisions regulate the assessment of water resources quantity and quality according to sectoral water uses, and define the methodology for assessment of the environmental flow to ensure safe ecological condition of water resources. However, the defined criteria for the flow relate more to hydrological than to environmental flow, whereas the fundamental concept in environmental flow is the recognition that water quality and quantity are intimately related. To address this gap, it might be useful to take the European Union's Common Implementation Strategy for the Water Framework Directive guidance document on developing ecological flow, which is in the process of development and will be completed in 2014, and will define a common definition and methodology for calculation of the environmental flow. The guidance document will be implemented by European Union member countries in the next cycle of RBMPs due for adoption by the end of 2015.
Development of the State Water Cadastre	11. Development and implementation of a short-term program for the State Water Cadastre, including development and introduction of the water resources coding system, establishment and introduction of modern maintenance mechanisms for the State Water Cadastre, and development of a water resources coding system.	State budget funding was provided in 2007–10 for implementation of a short-term program on the operation of the State Water Cadastre, including development of geographic information system (GIS) layers based on the State Water Cadastre, which is currently being used as a tool for data maintenance and provision and river basin characterization through coding of water resources. To support the SWCIS, works have been undertaken within the USAID-funded Program for Institutional and Regulatory Strengthening of Water Management in Armenia, 2005–09. The initiative was targeted toward participation of the main water stakeholder agencies in development of the State Water Cadastre database, and issues related to data provision and data exchange.	Despite significant progress in this direction the expected result of this activity—a centralized data warehouse accessible to the public—has not been achieved. Moreover, due to inadequate funding, the State Water Cadastre currently does not have electronic links to databases containing the initial data, as originally envisioned. Significant work needs to be conducted for strengthening the SWCIS, which cannot be completed without adequate funding. One of the options to improve the situation could be changing the status of the Division of Water Resources and making it a State noncommercial organization under the WRMA. This would give it the right to generate some income, which could then be used for strengthening the SWCIS.

table continues next page

Table D.1 Status of Implementation of the Short-Term Measures of the National Water Program (*continued*)

Issue	Short-term measure	Implementation status	Challenges and opportunities
Improvement of public awareness and participation in the water sector management process	12. Ensuring public awareness and participation in the planning and management of water resources at the national and basin management levels.	Not completed.	The European Union Water Framework Directive Common Implementation Strategy guidance document on public participation, providing decision makers and authorities with guidance on how to use public participation and stakeholder involvement to improve water management, can be used to implement this measure.
	13. Development and implementation of strategies for establishment of basin public councils, and building the technical capacity of the basin public councils.	Not completed.	Establishment of basin public councils requires corresponding funding, whereas currently even the staff of BMOs has been reduced following the economic crisis of 2008–09, due to budgetary constraints. Establishment of basin public councils does not therefore seem to be achievable at the moment and will become relevant only when BMOs are sufficiently developed and strengthened.
Implementation and monitoring of the National Water Program	14. Implementation and continuous monitoring and assessment of the National Water Program.	Not completed.	Despite that fact that at the moment there is no designated agency in charge of continuous monitoring of the implementation of the National Water Program, there are some opportunities with the recently established Water Resources Policy Division of the Ministry of Nature Protection.
			The main tasks of the division include formation of State policy on water resources protection, development of policy programs and strategic directions, and monitoring of their implementation. One of the specific functions of the division is development of the National Water Policy and monitoring of its implementation, though nothing is mentioned specifically about the National Water Program. However, given the general task of the division on monitoring the implementation of policy programs and strategic directions, it is ideally positioned to implement continuous monitoring of implementation of the National Water Program.

table continues next page

Table D.1 Status of Implementation of the Short-Term Measures of the National Water Program *(continued)*

Issue	Short-term measure	Implementation status	Challenges and opportunities
	15. Establishment of a monitoring system for the program implementation.	Not completed.	When the Water Resources Policy Division is officially designated as being in charge of implementation of continuous monitoring and assessment of the National Water Program, it can establish a sustainable monitoring system.
Development of plans for IWRM	16. Capacity building in the WRMA and basin management organizations for IWRM.	Several study tours to European countries and training courses on IWRM were organized with the support of projects funded by the European Commission, GEF, USAID, UNDP, and other donors. With the support of UNDP/GEF, a curriculum on Reducing Transboundary Degradation in the Kura-Araks River Basin is being developed to introduce an IWRM masters course at the Yerevan State University of Architecture and Construction.	One of the main challenges in this regard is that quite often the staff trained in IWRM techniques leave the State institutions because of uncompetitive salaries.
	17. Development of a pilot RBMP and identification of information needs for one basin management area.	With the support of USAID, European Union, GEF, UNDP, UNECE, and other donors, draft RBMPs have been developed in the Marmarik, Meghriget, Debed, Aghstev, Vorotan, and Arpa River basins. Currently draft RBMPs are being developed for the Akhuryan and Metsamor (Sevjur) River basins. Most of the plans have identified significant data gaps, which need to be filled for development of RBMPs that are fully compliant with the European Union Water Framework Directive. Within the European Union Water Initiative, based on the example of the Marmarik River basin, the government has drafted a Resolution on Approving the Contents of the Water Basin Management Model Plan, which was approved by Decision 4 of the government of Armenia Protocol Session of February 3, 2011. It will become the basis for development of the technical characteristics of the six RBMPs.	The capacity in the country to develop RBMPs is low. Implementation of existing plans is problematic. The main reason that the government of Armenia has not officially adopted existing draft RBMPs is the lack of appropriate funding to implement the planning recommendations. To address the problem, it is necessary to conduct cost-benefit analysis on the recommendations. This will enable the government of Armenia to see both the costs and economic benefits.

table continues next page

Table D.1 Status of Implementation of the Short-Term Measures of the National Water Program *(continued)*

Issue	Short-term measure	Implementation status	Challenges and opportunities
Implementation of the Lake Sevan Action Plan	18. Review and improvement of the annual and complex programs of measures for restoration, protection, reproduction, and use of the Lake Sevan ecosystem.	On May 15, 2001, the Law on Lake Sevan was adopted, which regulates the protection, regeneration, restoration, sustainable development, and use of Lake Sevan, its watershed, and the adjacent economic zone. On December 27, 2001, the Law on Adoption of the Annual and Complex Programs of Activities for the Use, Protection, Reconstruction, and Reproduction of the Lake Sevan Ecosystem was passed. The law sets annual (2002) and complex (from 2003 to 2030) measures to prevent decrease of the lake's level, increase it by 6 meters by 2030, protect the quality of the lake, ensure reproduction of fish stock, develop sustainable recreation and tourism, and improve the management of the lake.	In 2008 a Presidential Commission on Lake Sevan Issues was established, which among other things monitors the implementation of the complex program of measures.
Assessment of water resources and water reserve components	19. Clarification of up-to-date characteristics of water resources and water reserve components.		
	a) Clarification of quantitative and qualitative space-time characteristics of the surface natural renewable flow, taking into account quantitative and qualitative monitoring data, climate change, and anthropogenic impacts on the flow, accumulated during the last four decades.	Within the framework of the medium-term expenditure framework, a State budget-funded project was implemented in 2009 in some provinces of Armenia. Continued State budget funding is required to complete the task.	The European Environment Agency and the European Commission have developed methodologies for calculation of water accounts at river basin and subcatchment level. The intent is to inform water managers on how much water flows in and out of a river basin and how much realistically can be expected to be available before allocation takes place. Application of these methodologies, together with the use of hydrological models (such as the Water Evaluation and Planning system) can help to conduct separate assessment of surface water and groundwater reserves in Armenia.
	b) Adjustment of separate assessment of surface water and groundwater reserves, adjustment of deep flow data.	Not completed.	

table continues next page

Table D.1 Status of Implementation of the Short-Term Measures of the National Water Program *(continued)*

Issue	Short-term measure	Implementation status	Challenges and opportunities
	c) Development of methods for determination and calculation of the minimal ecological flow in main rivers.	Resolution 927-N of June 30, 2011, on Defining Drinking-Household and Agricultural Water Demand and Assessing the Environmental Flow According to River Basin Management Area, includes a method for determination and calculation of flow, but in reality it is not ecological, but rather sanitary or hydrological flow, and does not include any requirements on water quality.	To address this gap, it might be useful to apply the Common Implementation Strategy for the EU Water Framework guidance document on developing ecological flow, which will be finalized in 2014 and will define a common definition and methodology for calculation of environmental flow. The European Union member countries will implement the guidance document in the next cycle of RBMPs due for adoption by the end of 2015.
	d) Adjustment of the data on operating reserves of groundwater resources and definition of the permissible (maximum) water extractions.	Not completed. By the request of the government of Armenia, the USAID Clean Energy and Water Program currently supports implementation of a comprehensive assessment study of the groundwater resources of the Ararat valley, by using all the existing data and information from previously conducted assessment studies, available reports, and other documents.	Definition of exploitable groundwater reserves for the entire country is quite a challenging task, given that monitoring of groundwater resources stopped for about two decades. Though in 2009 the government of Armenia revitalized its Groundwater Monitoring Program, the unprocessed monitoring data for 2009–13 are not sufficient for assessing the resources and determining the sustainable rates for using groundwater resources.
Elaboration and enforcement of water quality standards	20. Adjustment and introduction of an internationally accepted methodology for determination of norms for the limitation of impacts on water resources and standards for ensuring water quality, taking into consideration best international practices.	The government adopted Resolution 75-N on Defining Water Quality Norms for Each Water Basin Management Area Taking into Consideration the Peculiarities of the Locality on January 27, 2011, which is based on internationally accepted methodologies and has been assessed as a relatively progressive document.	While the newly adopted surface water quality norms have been assessed as quite advanced and comprehensive by many international organizations, they still need some improvement. Particularly, the surface water quality norms for the Araks transboundary river and for the lakes and reservoirs of the country are not yet defined. In addition, there is a need to develop internationally acceptable water quality standards for groundwater resources.

table continues next page

Table D.1 Status of Implementation of the Short-Term Measures of the National Water Program *(continued)*

Issue	Short-term measure	Implementation status	Challenges and opportunities
Determination and conservation of aquatic ecosystem protection zones	21. Development of a methodology for determination of aquatic ecosystem protection zones, including flow formation zones, groundwater protection zones, water protection zones, ecotone, zones of ecological emergency and ecological disasters in water basins, and zones not subject to alienation.	The government developed and approved Resolution 64-N of January 20, 2005, on Criteria for Defining Water Ecosystem Sanitary Protection Zone, Territories of Groundwater Protection, Water Protection, Ecotone, and Nondisposable Zones.	Implementation of the resolution has proved problematic. Only some of the major drinking water intake structures have sanitary protection zones. Thus, the overall objective of this activity, which is to protect surface water and groundwater resources from pollution, has not yet been achieved.
Study of the status of previously drained agricultural lands in the Ararat valley	22. Development and implementation of programs for use of previously drained agricultural lands in the Ararat valley.	Rehabilitation of the Ararat valley drainage system was conducted in 2010–12 within the irrigation infrastructure activity of the Millennium Challenge Account Armenia Program. Prior to rehabilitation works, an environmental and hydrological baseline study of wetlands in the Ararat valley was conducted. Rehabilitation of the drainage system included removal of old blockages in the drains and provision of substitute measures to provide irrigation water, cleaning and deepening of collectors and secondary and tertiary drains, improvement of the drainage system by constructing culverts and numerous small structures, rehabilitation of a number of artesian wells, and a limited amount of work on subsurface drainage, while maintaining an optimally balanced ecosystem in the wetlands. As a result, the groundwater level is now 2 meters lower, and it is already possible to cultivate higher value crops.	—

table continues next page

Table D.1 Status of Implementation of the Short-Term Measures of the National Water Program *(continued)*

Issue	Short-term measure	Implementation status	Challenges and opportunities
Increasing the strategic water reserves and regulation of river flow	23. Implementation of works provided for under the program for reservoir construction.	With the funding of KfW, the selected consultant in 2013 started the feasibility study for construction of Kaps reservoir and a gravity irrigation system. In the initial phase it is anticipated to rehabilitate the dam at a low level, providing about 6 million cubic meters of capacity and the gravity supply of irrigation water for 2,200 hectares currently supplied by pumps or not irrigated. With the funding of AFD, a feasibility study started in 2013 for construction of the Vedi reservoir with an overall storage volume of 20 million cubic meters (dam height—70 meters). Construction of the reservoir will make it possible to irrigate 2,744 hectares of agricultural lands and remove water intake structures. In the territory of the Myasnikyan community of Armavir province, on the Mastara River, it is planned to construct the Mastara reservoir with 10.2 million cubic meters overall storage volume (planned height of the dam—30 meters). It will collect the free flow of the Selav-Mastara River in the section between the Akhuryan reservoir and the head structure of the Talin irrigation system. The construction of the reservoir will enable irrigation of 4,384 hectares of agricultural lands in the region. As potential funder, the KF is planning to conduct an appraisal mission. Also, the government of Armenia has applied to JICA for the construction of Yegvard reservoir with 90 million cubic meters overall storage volume (planned height of the dam—32 meters).	—

table continues next page

Table D.1 Status of Implementation of the Short-Term Measures of the National Water Program *(continued)*

Issue	Short-term measure	Implementation status	Challenges and opportunities
Implementation of water quality management	24. Development of a strategy for water quality management.	The government adopted Resolution 75-N on Defining Water Quality Norms for Each Water Basin Management Area Taking into Consideration the Peculiarities of the Locality on January 27, 2011. The allowable limits of potential pollutants, impacting the surface water quality, are defined for all six basin management areas, taking into consideration the peculiarities of the locality. Water quality norms are defined for all river basins, taking account of the requirements of the European Union Water Framework Directive, as well as the hydromorphological, hydrogeographical, hydrophysical, environmental, and other peculiarities of the country. At the same time, within the framework of the European Union Water Initiative National Policy Dialogues in Armenia, a payment for ecosystem services scheme has been introduced for the upper Hrazdan River basin, the first time such a scheme has been implemented in Armenia. According to the scheme, it is proposed to set a limit for units of pollutants, taking into consideration the polluter pays principle.	In order to develop a strategy for water quality management, there is a need to establish a system for water quality assessment. To accomplish that task, currently the Ministry of Nature Protection is in the process of establishing an interagency working group on water quality monitoring and water quality assessment, which in the next two to three years should work on transition toward a water quality monitoring and water quality assessment system that is compliant with the European Union Water Framework Directive.
Development of spatial planning criteria and guidelines	25. Review and improvement of the existing approaches to spatial planning.	Not completed.	Spatial planning is key for integration of water management and land use planning in river basin districts. This is particularly important for Armenia, taking into consideration the physical and spatial characteristics of the rivers. However, there is rather weak coordination between the spatial planning, land use planning, and water resource management systems.

table continues next page

Table D.1 Status of Implementation of the Short-Term Measures of the National Water Program *(continued)*

Issue	Short-term measure	Implementation status	Challenges and opportunities
Management of transboundary water resources	26. Development of a program for management of transboundary water resources.	No program has been developed, but Armenia is actively participating in all water-related transboundary projects and initiatives in the region. Formally, Armenia has several international agreements in place with its neighboring countries on transboundary water resources (with Turkey, signed in 1927, 1964, 1973, 1975, 1990; with Georgia, signed in 1971, 1997; with Azerbaijan, signed in 1974; and with the Islamic Republic of Iran, signed in 1957, 2006). However, only joint monitoring and measurement activities take place with the Islamic Republic of Iran and Turkey.	In 2002 the Armenian Commission on Transboundary Water Resources was established by a decision of the Prime Minister. The basic functions of the commission include (a) formulation of draft inter-State agreements and their submission to the government; (b) notification to the relevant agencies of issues not regulated by inter-State agreements and requiring due resolution; and (c) provision of information to agencies in Armenia concerning the state of transboundary waters and transboundary impacts. However, the commission does not have any support staff or secretariat and since its establishment practically has not implemented any activity.
Water systems management needs			
Improvement of water supply and wastewater collection services	27. Study of water supply and wastewater collection services (drinking-domestic water supply, irrigation, hydropower generation, etc.) and development and implementation of programs aimed at improvement of the provided services.	Several studies have been completed by international donor organizations and international financial institutions, including JICA, World Bank, ADB, OECD, and EBRD, coordinated by the State Committee on Water Systems. The results of such studies have been used by the international operators of water supply and wastewater collection services. In the hydropower generation sector, in 2008 a comprehensive report—Update of the Existing Scheme for Small Hydropower Stations of the Republic of Armenia—was prepared with the funding of GEF. Within the European Union Water Initiative, in 2013 OECD started work on development of a wastewater collection and treatment strategy in Armenia.	Despite achievements in this direction in the last decade, progress lags for the self-supplied rural communities that are outside the service area of water supply and sanitation companies.

table continues next page

Table D.1 Status of Implementation of the Short-Term Measures of the National Water Program (*continued*)

Issue	Short-term measure	Implementation status	Challenges and opportunities
Ensuring safety of hydrotechnical structures	28. Development of programs aimed at enhancing the effectiveness of measures for ensuring the safety of hydrotechnical structures and reliability of operations.	Together with the Abu Dhabi Foundation, the government of Armenia has co-funded the Arpa-Sevan tunnel rehabilitation works and safety measures. Within the World Bank Dam Safety Project II, rehabilitation and safety improvement works have been conducted for 44 dams. Also, corresponding technical assistance, training, and equipment support has been provided to water supply agencies and the Ministry of Emergency Situations, enhancing their response capacity for dealing with emergency cases. Regulations and necessary procedures for dam operation, training of operators, and provision of heavy equipment for operation and maintenance of dams were prepared.	Adequate and regular maintenance of hydrotechnical structures requires sufficient institutional capacity and financial resources. Although the government of Armenia is committed to supporting those, budget constraints affect provision of adequate funding to the operation and maintenance needs of hydrotechnical structures. According to the World Bank Dam Safety Project completion report, there is a need to continue to optimize the operation and use of water stored behind the dams, especially for those that are built as a cascade on rivers.
	29. Clarification of responsibilities for operation and protection of hydrotechnical structures of State significance.	Joint measures have been implemented by the Ministry of Emergency Situations and the Dam Operation Department of the State Committee on Water Systems.	–

European Union Water Framework Directive

Directive 2000/60/EC of the European Parliament and of the Council establishing a framework for the Community action in the field of water policy—or, in short, the EU Water Framework Directive—was adopted on October 23, 2000. The Water Framework Directive commits European Union Member States to making all water bodies (including marine waters) of good qualitative and quantitative status by 2015. It is a framework that prescribes steps to reach the common goal rather than adopting the more traditional limit value approach. Good ecological status is defined locally as being lower than a theoretical reference point of pristine conditions, that is, no anthropogenic influence.

The Water Framework Directive includes 25 articles and 10 annexes, as follows (European Union 2000):

Article 1   Purpose
Article 2   Definitions
Article 3   Coordination of administrative arrangements within river basin districts
Article 4   Environmental objectives
Article 5   Characteristics of the river basin district, review of the environmental impact of human activity, and economic analysis of water use
Article 6   Register of protected areas
Article 7   Water used for the abstraction of drinking water
Article 8   Monitoring of surface water status, groundwater status, and protected areas
Article 9   Recovery of costs for water services
Article 10   The combined approach for point and diffuse sources
Article 11   Program of measures
Article 12   Issues that cannot be dealt with at Member State level

Annex V explains three areas of water quality monitoring for surface water and groundwater—surveillance, operational, and investigative monitoring, as elaborated below.

Monitoring of Ecological Status and Chemical Status for Surface Waters

Member States shall, for each period to which a river basin management plan applies, establish a surveillance monitoring program and an operational monitoring program. Member States may also need in some cases to establish programs of investigative monitoring.

Member States shall establish **surveillance monitoring** programs to provide information for:

- Supplementing and validating the impact assessment procedure detailed in Annex II, and the efficient and effective design of future monitoring programs;
- Assessment of long-term changes in natural conditions;
- Assessment of long-term changes resulting from widespread anthropogenic activity.

Operational monitoring shall be undertaken in order to:

- Establish the status of those bodies identified as being at risk of failing to meet their environmental objectives;
- Assess any changes in the status of such bodies resulting from the program of measures.

Investigative monitoring shall be carried out:

- Where the reason for any exceedance is unknown;
- Where surveillance monitoring indicates that the objectives set out in Article 4 for a body of water are not likely to be achieved and operational monitoring has not already been established, in order to ascertain the causes of a water body or water bodies failing to achieve the environmental objectives;
- To ascertain the magnitude and impacts of accidental pollution.

Monitoring of Chemical Status for Groundwater

The monitoring network shall be designed so as to provide a coherent and comprehensive overview of groundwater chemical status within each river basin and to detect the presence of long-term anthropogenically induced upward trends in pollutants. On the basis of the characterization and impact assessment carried out in accordance with Article 5 and Annex II, Member States shall, for each period to which a river basin management plan applies, establish a surveillance monitoring program. The results of this program shall be used to establish an operational monitoring program to be applied for the remaining period of the plan.

Surveillance monitoring shall be carried out in order to:

- Supplement and validate the impact assessment procedure;
- Provide information for use in the assessment of long-term trends both as a result of changes in natural conditions and through anthropogenic activity.

Operational monitoring shall be undertaken in the periods between surveillance monitoring programs in order to:

- Establish the chemical status of all groundwater bodies or groups of bodies determined as being at risk;
- Establish the presence of any long-term anthropogenically induced upward trend in the concentration of any pollutant.

Reference

European Union. 2000. *Directive 2000/60/EC of the European Parliament and of the Council of 23 October 2000 Establishing a Framework for Community Action in the Field of Water Policy*. Water Framework Directive.

Fish Farms with Large Water Uses

Table F.1 Fish Farms with Water Intake Larger than 300 L/s (9.5 MCM/yr) in Ararat Valley

Water user	Business address (marz/village)	Number of wells with WP	WP volume (L/s)	Actual intake as of 01.07.2013 (L/s)	Water intake over WP or without WP (L/s)
1	Ararat/Ranchpar	23	3,054	4,500 (142 MCM/yr or 13%[a])	1,446
2	Ararat/Sayat-Nova	20	2,706	3,600 (114 MCM/yr or 10%[a])	894
3	Ararat/Sayat-Nova	11	1,647.5	1,700 (54 MCM/yr or 5%[a])	52.5
4	Ararat/Sis	10	1,495	950	−545
5	Ararat/Zorak	6	606	828	222
6	Ararat/Sayat-Nova	10	1,169	750	−419
7	Ararat/Ranchpar	4	600	720	120
8	Ararat/Hovtashat	4	443	700	257
9	Ararat/Marmarashen	6	648	675 (545 with 6 wells with WP and 130 from 2 wells without WP)	27
10	Ararat/Sayat-Nova	7	640	663	23
11	Ararat/Sipanik	4	710	640	−70
12	Ararat/Noramarg	5	558	550	−8
13	Ararat/Hayanist	4	555	550	−5
14	Armavir/Gay	7	555	490	−65
15	Armavir/Gay	8	852	460	−392
16	Ararat/Ranchpar	3	450	460	10
17	Ararat/Sipanik	6	900	400	−500
18	Ararat/Noramarg	2	400	350	−50
19	Ararat/Sipanik	2	300	330	30
20	Armavir/Gay	6	507	320	−187
21	Armavir/Araks	6	213	310	97
22	Armavir/Gay	8	282	310	28
Total				**20,256 (639 MCM/yr or 57%[a])**	**9,021 (284 MCM/yr or 25%[a])**

Source: USAID 2013.
Note: L/s = liters per second; MCM = million cubic meters a. Percentage in total water intake by all fish farms in Ararat valley (35,497.3 L/s or 1,119 MCM/yr).

Reference

USAID (United States Agency for International Development). 2013. *Analysis and Assessment of Groundwater in Ararat Valley*. Interim reports 1 and 2, prepared under USAID Clean Energy and Water Program.

Transboundary Surface Water Flows

Table G.1 Transboundary Surface Water Flows

Macro basin	River basin	Surface area (km² and percentage)					Transboundary flows (BCM)	
		Armenia	Georgia	Azerbaijan	Turkey	Iran, Islamic Rep.	Inflows	Outflows
Araks	Araks	22,560	0	18,140	19,500	41,800	0	5.01
		22%	0%	18%	19%	41%		
Araks	Akhuryan	2,784	0	0	6,916	0	2.12	0
		29%	0%	0%	71%	0%		
Kura	Aghstev	770	0	1,730	0	0	0	0.29
		31%	0%	69%	0%	0%		
Kura	Debed	3,790	310	0	0	0	0	1.04
		92%	8%	0%	0%	0%		
Araks	Arpa	2,080	0	550	0	0	0	0.53
		79%	0%	21%	0%	0%		
Araks	Vorotan	2,030	0	3,620	0	0	0	0.69
		36%	0%	64%	0%	0%		
Araks	Voghji	788	0	387	0	0	0	0.37
		67%	0%	33%	0%	0%		
						Total	**2.12**	**7.93**

Source: Adapted from Hannan, Leummens, and Matthews 2013.
Note: Transboundary flow into the Akhuryan basin is net from the flow inside Armenia (0.39 MCM).
BCM = billion cubic meters; MCM = million cubic meters.

Reference

Hannan, T., H. J. L. Leummens, and M. M. Matthews. 2013. *Desk Study: Hydrology.* UNDP/GEF Reducing Transboundary Degradation in the Kura Araks River Basin Project.

Existing Agreements for Transboundary Cooperation in the Watersheds of the Kura-Araks River Basin

Table H.1 Existing Agreements for Transboundary Cooperation in the Watersheds of the Kura-Araks River Basin

Countries	Watershed	Title	Date signed (S) or entry into force (E)
Armenia and Turkey	Araks	Convention between the Republic of Turkey and the Union of Soviet Socialist Republics concerning Water Use of Transboundary Waters	1927 (S)
		The Bilateral Commission between Armenia and Turkey operates on this basis	1928 (E)
Armenia and Turkey	Akhuryan/ Arpacay	Protocol Concerning Mainly Technical Cooperation, Riverbed Changes, and Construction of Joint Hydrotechnical Facilities, extending between border stone number 41 through border stone number 450 on the Turkish-Soviet Union border	1990
Armenia and Turkey	Akhuryan/ Arpacay	The Protocol of the Meeting of the Turkish-Soviet Joint Commission Pertaining to the Joint Construction of a Dam on the Arpacay (Akhuryan)	1964
Armenia and Turkey	Akhuryan/ Arpacay	Agreement between the Republic of Turkey and the Union of Soviet Socialist Republics on the Joint Exploitation of Dam and Reservoir on the Akhuryan (Arpachay) River	1973
Armenia and Turkey	Akhuryan/ Arpacay	Cooperation Agreement between the Republic of Turkey and the Union of Soviet Socialist Republics on the Construction of a Dam on the Bordering Arpacay (Akhuryan) River and the Constitution of a Dam Lake	1975
Armenia and Azerbaijan	Vorotan/ Bargushad	Agreement between the Soviet Socialist Republic of Armenia and the Soviet Socialist Republic of Azerbaijan on the Joint Utilization of the Waters of the River Vorotan	1974

table continues next page

Table H.1 **Existing Agreements for Transboundary Cooperation in the Watersheds of the Kura-Araks River Basin** *(continued)*

Countries	Watershed	Title	Date signed (S) or entry into force (E)
Armenia and Iran, Islamic Rep.	Araks	Treaty between the Government of the Union of Soviet Socialist Republics and the Imperial Government of Iran Concerning the Regime of the Soviet Iranian Frontier and the Procedure for the Settlement of Frontier Disputes and Incidents	1957 (S)
		The Bilateral Commission between Armenia and the Islamic Republic of Iran acts on this basis	1957 (S)
Armenia and Iran, Islamic Rep.	Araks	Agreement between Iran and the Union of Soviet Socialist Republics for the Joint Utilization of the Frontier Parts of the Rivers Araks and Atrak for Irrigation and Power Generation and Domestic Use	1957 (S)
Armenia and Georgia	No specific watershed (general framework)	Agreement between the Governments of Georgia and of the Republic of Armenia on Cooperation in Environmental Protection	1997 (S)
Armenia and Georgia	Debed	Protocol of Agreement between Armenia and Georgia on the Design of a Water Intake in the Debed River	1971 (S)
Armenia and Azerbaijan	Arpa	Agreement between the Council of Ministers of the USSR and the Council of Ministers of the Azerbaijan SSR on Transfer of Arpa River into Lake Sevan	1962 (S)

Source: UNECE 2009.

Reference

UNECE (United Nations Economic Commission for Europe). 2009. *River Basin Commissions and Other Institutions for Transboundary Water Cooperation.* Capacity for Water Cooperation in Eastern Europe, Caucasus, and Central Asia. Geneva: UNECE.

APPENDIX I

Status of Ratification of Multilateral Treaties by Armenia and Its Neighbors

Table I.1 Status of Ratification of Multilateral Treaties and Customary International Law by Armenia and Its Neighbors

Treaty	Armenia	Azerbaijan	Georgia	Iran, Islamic Rep.	Turkey
Convention on Protection and Use of Transboundary Watercourses and International Lakes March 17, 1992, Helsinki, Finland	–	Party	–	–	–
Protocol on Water and Health (to the Transboundary Watercourses Convention) June 17, 1999, London, United Kingdom	Signatory	Party	Signatory	–	–
Protocol on Civil Liability and Compensation for Damage Caused by Transboundary Effects (to the Transboundary Watercourses Convention) May 21, 2003, Kiev, Ukraine	Signatory	–	Signatory	–	–
Convention on Environmental Impact Assessment in a Transboundary Context February 25, 1991, Espoo, Finland	Party	Party	–	–	–
Protocol on Strategic Environmental Assessment (to the Convention on Environmental Impact Assessment in a Transboundary Context) May 21, 2003, Kiev, Ukraine	Party	–	Signatory	–	–
Convention on the Transboundary Effects of Industrial Accidents 1992, Helsinki, Finland	Party	Party	–	–	–

table continues next page

Table I.1   Status of Ratification of Multilateral Treaties and Customary International Law by Armenia and Its Neighbors (continued)

Treaty	Armenia	Azerbaijan	Georgia	Iran, Islamic Rep.	Turkey
Convention on Wetlands of International Importance especially as Waterfowl Habitat February 2, 1971, Ramsar, Iran, Islamic Rep.	Party	Party	Party	Party	Party
Convention on Biological Diversity June 5, 1992, Rio de Janeiro, Brazil	Party	Party	Party	Party	Party

Source: UNECE 2011.

Note: Table includes signatory, ratification, party. – represents no participation/action. The Islamic Republic of Iran is outside the UNECE region, but the entry into force of articles 25 and 26 will allow accession by countries outside this region.

Convention/Protocol websites:

Transboundary	http://www.unece.org/
Water and Health	http://www.unece.org/
Wetlands	http://www.ramsar.org/
Biological Diversity	http://www.biodiv.org/

Reference

UNECE (United Nations Economic Commission for Europe). 2011. *Second Assessment of Transboundary Rivers, Lakes, and Groundwaters.* Convention on the Protection and Use of Transboundary Watercourses and International Lakes.

Details of World Bank Water-Related Activities in Armenia

World Bank water-related activities are shown in tables J.1 and J.2. There are projects across many water-related sectors, including providing physical infrastructure upgrades and technical assistance for institutional and regulatory reforms. In addition, the World Bank-led or -supported preparation of an IWRM planning study, the National Environmental Action Plan, the Lake Sevan Action Program, and the Poverty Reduction Strategy Papers, which identified challenges and priority actions for sustainable water resources management in Armenia.

Table J.1 Ongoing or Planned World Bank Projects

Theme	Project	Implementation
Water supply and wastewater management	Municipal Water Project (P126722)	Feb 2012–present
	Water Tariff Study (P146342): TA	FY15
Irrigation	Irrigation System Enhancement Project (P127759)	May 2013–present
	Irrigation System Modernization Project (P147310): TA	FY15
Environment (mining industry)	Armenia Environment Sector Note (P132911): TA	FY14
Solid waste management	Transaction Advisory Support for Public-Private Partnerships for Solid Waste Management for Yerevan City (P118936): TA	—

Note: — = not available.

Table J.2 Closed World Bank Projects

Theme	Project	Implementation	Outcomes rating
Water supply and wastewater management	Yerevan Water and Wastewater Project (P087641)	Sep 2005–Dec 2011	Satisfactory
	Municipal Water and Wastewater Project (P063398)	Nov 2004–Feb 2012	Moderately satisfactory
Hydropower	Renewable Energy Project (P083352)	Aug 2006–Jun 2011	Satisfactory
Watershed management	Natural Resources Management and Poverty Reduction Project (P057847)	Dec 2002–Jan 2009	Moderately satisfactory
Dam safety	Irrigation Dam Safety 2 Project (P088499)	Dec 2004–Dec 2009	Satisfactory
	Dam Safety Project (P064879)	Apr 2000–Sep 2009	Highly satisfactory
Irrigation	Irrigation and Rehabilitation Emergency Project (P116681)	Oct 2009–Jun 2013	Satisfactory
	Irrigation Development Project (P055022)	Dec 2001–Mar 2009	Satisfactory
	Irrigation Rehabilitation Project (P008277)	Mar 1995–May 2001	Satisfactory

Source: World Bank 2001, 2009a, 2009b, 2010a, 2010b, 2012a, 2012b, 2012c, 2013.

References

World Bank. 2001. *Irrigation Rehabilitation Project: Implementation Completion and Results Report.* Report No. 23168.

———. 2009a. *Irrigation Development Project: Implementation Completion and Results Report.* Report No. ICR00001145.

———. 2009b. *Natural Resources Management and Poverty Reduction Project: Implementation Completion and Results Report.* Report No. ICR00001040.

———. 2010a. *Armenia Dam Safety Project: Implementation Completion and Results Report.* Report No. ICR00001144.

———. 2010b. *Irrigation Dam Safety II Project: Implementation Completion and Results Report.* Report No. ICR00001428.

———. 2012a. *Municipal Water and Wastewater Project: Implementation Completion and Results Report.* Report No. ICR2347.

———. 2012b. *Renewable Energy Project: Implementation Completion and Results Report.* Report No. ICR1960.

———. 2012c. *Yerevan Water and Wastewater Project: Implementation Completion and Results Report.* Report No. ICR2346.

———. 2013. *Irrigation Rehabilitation Emergency Project: Implementation Completion and Results Report.* Report No. ICR00002815.

Other International Donor Water-Related Activities

International Finance Corporation (IFC). Armenia became a member and shareholder of the IFC in 1995. The IFC began providing advisory services in Armenia in 1999 and investing in 2000. The IFC financed the three-year (2010–13) Armenia Sustainable Energy Finance Project. This project was designed to establish a sustainable market for investments in energy efficiency and renewable energy, including hydropower.

Asian Development Bank (ADB). Since Armenia became an ADB member country in 2005, the ADB has supported water supply and sanitation, rehabilitation of the Sevan-Hrazdan cascade hydropower system, and solid waste management.

Eurasian Development Bank (EDB). Armenia became an EDB member country in 2009. The EDB is currently preparing a project on rehabilitation and modernization of irrigation systems. This investment will complement the ongoing World Bank irrigation project.

European Bank for Reconstruction and Development (EBRD). The EBRD has worked with Armenia since 1992 and is the largest investor in the private enterprise and financial sectors of Armenia. The EBRD has supported various projects for water supply in Yerevan and small municipalities, Lake Sevan protection through wastewater management, solid waste management, and hydropower rehabilitation. EBRD priorities for 2012–15 include continued support for the municipal and environmental infrastructure sectors and sustainable energy development.

European Union/European Commission (EU/EC). The EU/EC implemented projects on transboundary river management for the Kura River (Armenia, Azerbaijan, and Georgia) during 2002–04 and 2008–13. These projects supported the development of a common monitoring and information management system to improve transboundary cooperation in the Kura River basin. Draft

Aghstev and Debed RBMPs have been prepared, based on EU Water Framework Directive requirements. Armenia is also a priority partner country within the European Neighborhood Policy. A joint European Union-Armenia Action Plan, which includes regional cooperation for water resources management, was adopted in 2006. As a part of this plan, the European Neighborhood Partnership Instrument–Shared Environmental Information System (ENPI-SEIS) was adopted by Armenia. In 2012, another project was launched with the European Union (until 2016) the Environmental Protection of International River Basins Project, which includes protection of the Kura River basin. The project is currently drafting the Akhuryan-Metsamor RBMP.

Organization for Security and Co-operation in Europe (OSCE). The OSCE established a Yerevan office in 1999, and it started operating in 2000. The OSCE has assisted the Armenian government in setting up and operating the Aarhus Sustainability Network, a public environmental information center for public participation and regional cooperation. The Environment and Security Initiative (a partnership of OSCE, UNEP, UNECE, UNDP, REC, and NATO) includes projects for transboundary water, environmental protection, and climate change studies for the south Caucasus region. The OSCE is providing support to Armenia on this.

United Nations Economic Commission for Europe (UNECE). The main operation instrument of the European Union Water Initiative in Armenia, which started in 2006 and will continue until 2015, is the National Policy Dialogue on IWRM and on water supply and sanitation. UNECE is the strategic partner for support to the policy dialogue process on IWRM. The following projects were implemented in Armenia within the National Policy Dialogue process: a pilot RBMP in line with the principles of IWRM and the EU Water Framework Directive developed for the Marmarik River basin, a critical review of existing economic instruments for water management identifying recommendations for reforming existing instruments and introducing new instruments, the potential application of payment for ecosystem services (building on a pilot study in the upper Hrazdan River basin), and a program of actions under the UNECE Protocol on Water and Health. Finally, a strategy for wastewater collection and treatment in Armenia is nearing completion. It will provide specific recommendations on how to increase the coverage in a financially realistic way.

Organization for Economic Cooperation and Development (OECD). In the National Policy Dialogue of Armenia, OECD is the strategic partner for water supply and sanitation and economic and financial aspects of IWRM. It has conducted several financial analyses in the water supply and sanitation sector.

United Nations Development Program (UNDP). UNDP in Armenia was established in March 1993 and supports the government to reach national development priorities and the Millennium Development Goals by 2015. UNDP has been an active partner, particularly on climate change issues. UNDP supported the Armenian government in preparing the national communications to the

United Nations Framework Convention on Climate Change and assessing the socioeconomic impact of climate change in Armenia, including that on the water sector. UNDP supported the development of the first (in 1998) and second (in 2008) National Environmental Action Program of Armenia. In addition, UNDP supported a Lake Sevan coastal zone cleaning project in 2012–13 and implemented a "Revive a River" project to improve wastewater management for the Aghstev River in 2009–13. The ongoing projects supported by UNDP include disaster risk reduction and prevention and public-private partnerships in solid waste management. UNDP has also been implementing a regional project—Reducing Transboundary Degradation of the Kura-Araks River Basin—with financing from GEF and Sida. Armenia, Azerbaijan, and Georgia have participated in the first two phases of the regional project, and Armenia has not committed to continue on to the third phase.

United States Agency for International Development (USAID). The U.S. government has been the largest bilateral donor in Armenia, and USAID has been actively engaged with the water sector in the country. Since 2000, USAID has funded several national and regional water projects in Armenia, which has helped achieve measurable progress in the sector through both physical infrastructure upgrades and institutional and regulatory reforms. USAID supported IWRM in Armenia by providing software programs for river basin management and by developing and revising a model guideline for formulation of RBMPs. USAID was also engaged in developing the current water permit system and in strengthening the water monitoring capacity in the country (particularly with respect to groundwater monitoring). USAID provided modern laboratory equipment for water quality monitoring and developed the SWCIS. USAID has also supported water supply rehabilitation in targeted rural areas through small-scale infrastructure projects. Currently, USAID is supporting the preparation of RBMPs in the Vorotan, Megrhiget, and Voghji river basins. The ongoing Clean Energy and Water Program for 2011–15 includes activities to improve Armenia's energy and water security by developing sustainable hydropower, supporting water management, and integrated energy and water planning. Under the program, USAID completed a major comprehensive assessment of the growing fisheries and groundwater problem in the Ararat valley.

Millennium Challenge Corporation (MCC). The U.S. government's Millennium Challenge Corporation had a five-year compact with the government of Armenia to reduce rural poverty through improvements in the agricultural sector. The Millennium Challenge Corporation Armenia compact program, which closed in September 2011, included one of Armenia's largest-ever irrigation infrastructure projects. One of the activities included rehabilitation of the drainage system in the Ararat valley.

Japan International Cooperation Agency (JICA). JICA started working in the water sector of Armenia in 2007. In 2007–09 it financed the study for improvement of rural water supply and sewerage systems in the Republic of Armenia. Currently, JICA is considering a project to support the Yegvard reservoir.

Swedish International Development Cooperation Authority (Sida). Sida has provided support to transboundary issues and river basin planning in the country.

German Development Bank (KfW). This German government-owned development bank has an ongoing nationwide water supply project for improving local water supply facilities in the cities of Armavir, Metsamor, Gyumri, and Vanadzor, as well as surrounding villages and municipalities. The KfW renewable energy program provides long-term loans to promote private investments in the construction of new small hydropower plants. In addition, KfW is preparing a project on Kaps reservoir, and is conducting a feasibility study on the construction of the reservoir and associated gravity irrigation system. It will also support the preparation of a comprehensive plan for the Akhuryan River. The Kaps reservoir project will be a part of a larger IWRM program envisioned, which includes Lake Arpi protection and drinking water supply in the area.

German Agency for International Cooperation (GIZ). GIZ has an ongoing project on sustainable biodiversity management in the South Caucasus (Armenia, Azerbaijan, and Georgia) from 2008 to 2015.

French Development Agency (AFD). AFD is currently funding a feasibility study on the construction of Vedi reservoir.

Kuwait Fund for Arab Economic Development (KF). KF is considering a project on the Selav-Mastara reservoir.

Government of Norway (GoN). Several projects have been implemented in Armenia with financial support from the Norwegian Ministry of Foreign Affairs in the fields of environment, water, and energy. Some of the projects implemented include sustainable small hydropower development, biodiversity protection and eco-regional conservation planning (implemented with the WWF), and cooperation between the ASHMS and the Norwegian Water Resources and Energy Directorate in the field of operational hydrology. For hydropower, a new broad-based Norwegian-Armenian hydropower cooperation project was launched in Armenia in November 2011.

Environmental Benefits Statement

The World Bank Group is committed to reducing its environmental footprint. In support of this commitment, the Publishing and Knowledge Division leverages electronic publishing options and print-on-demand technology, which is located in regional hubs worldwide. Together, these initiatives enable print runs to be lowered and shipping distances decreased, resulting in reduced paper consumption, chemical use, greenhouse gas emissions, and waste.

The Publishing and Knowledge Division follows the recommended standards for paper use set by the Green Press Initiative. Whenever possible, books are printed on 50 percent to 100 percent postconsumer recycled paper, and at least 50 percent of the fiber in our book paper is either unbleached or bleached using Totally Chlorine Free (TCF), Processed Chlorine Free (PCF), or Enhanced Elemental Chlorine Free (EECF) processes.

More information about the Bank's environmental philosophy can be found at http://crinfo.worldbank.org/wbcrinfo/node/4.

green press INITIATIVE

www.ingramcontent.com/pod-product-compliance
Lightning Source LLC
Chambersburg PA
CBHW082356270326
41935CB00013B/1641